More Praise for
AT THE EDGE OF TIME

"What a journey, from the very birth of the universe to its ultimate future. In accessible fashion, Hooper's book does a great job explaining the fundamental laws of physics and showing how they play out in cosmic evolution."
—SEAN CARROLL, author of *Something Deeply Hidden*

"Where Weinberg's *The First Three Minutes* left off, Hooper's *At the Edge of Time* picks up. A riveting tour of modern cosmology told by one of its savviest guides, Hooper takes us on a journey from our universe's formerly inscrutable past to mesmerizing possible scenarios in its far future."
—BRIAN KEATING, author of *Losing the Nobel Prize*

"A clear and engaging tour of the mysterious birth of our universe, *At the Edge of Time* will keep you at the edge of your mental seat."
—DANIEL WHITESON, coauthor of *We Have No Idea*

"*At the Edge of Time* is a gripping tale of the monumental discoveries and unsolved mysteries in cosmology. Well-written and exciting, Hooper's book leads us from the early days of Einstein to the puzzles of the modern era, as well as through the author's own adventures in the search for answers."
—KATHERINE FREESE, author of *The Cosmic Cocktail*

"This book convincingly guides readers through some of the hottest topics in modern physics and astronomy: Big Bang theory, dark matter, dark energy, and gravitational waves. Bringing a fresh perspective, Hooper effectively captures the feelings of the community of scientists working to solve the greatest mysteries and demonstrates that a scientific revolution might be around the corner."
—GIANFRANCO BERTONE, author of *Behind the Scenes of the Universe*

T0007362

AT THE EDGE OF TIME

Books in the *Science Essentials* series bring cutting-edge science to a general audience. The series provides the foundation for a better understanding of the scientific and technical advances changing our world. In each volume, a prominent scientist—chosen by an advisory board of National Academy of Sciences members—conveys in clear prose the fundamental knowledge underlying a rapidly evolving field of scientific endeavor.

For a full list of titles in the series, go to https://press.princeton.edu /catalogs/series/title/science-essentials.html.

At the Edge of Time

EXPLORING THE
MYSTERIES OF OUR
UNIVERSE'S FIRST SECONDS

DAN HOOPER

PRINCETON UNIVERSITY PRESS

PRINCETON & OXFORD

Requests for permission to reproduce material from this work
should be sent to permissions@press.princeton.edu

Published by Princeton University Press
41 William Street, Princeton, New Jersey 08540
6 Oxford Street, Woodstock, Oxfordshire OX20 1TR
press.princeton.edu

First paperback printing, 2021
Paperback ISBN 978-0-691-20642-4

The Library of Congress has catalogued the cloth edition as follows:

Names: Hooper, Dan, 1976– author.
Title: At the edge of time : exploring the mysteries of our universe's first seconds /
Dan Hooper.
Description: Princeton : Princeton University Press, [2019] | Includes index.
Identifiers: LCCN 2019019248 | ISBN 9780691183565 (hardcover)
Subjects: LCSH: Cosmology—Popular works. | Big bang theory—Popular works.
Classification: LCC QB982 .H658 2019 | DDC 523.1—dc23
LC record available at https://lccn.loc.gov/2019019248

British Library Cataloging-in-Publication Data is available

Editorial: Jessica Yao and Arthur Werneck
Production Editorial: Natalie Baan
Production: Jacquie Poirier
Publicity: Sara Henning-Stout and Katie Lewis

Cover illustration and design by Sukutangan

This book has been composed in Arno Pro

Printed in the United States of America

For Cheryl

CONTENTS

ACKNOWLEDGMENTS

I'D LIKE TO THANK all of those who provided me with feedback about this book, especially Daniel Whiteson, Chris Vale, Becky Hooper, and Andy Mareska. I'd also like to thank my editor, Jessica Yao, for pushing me to make this book better than it otherwise would have been and providing me with great advice, even when I didn't want to hear it.

AT THE EDGE OF TIME

The Big Bang

Today

~2.7 K

Dark Energy Era Begins

~3.7 K

Solar System Forms

~3.8 K

Birth of the First Stars
(Cosmic Dawn)

~50 K

Formation of the First Atoms
(Origin of the Cosmic Microwave Background)

3000 K

Dark Matter Era Begins

~10,000 K

Formation of the First Nuclei
(Big Bang Nucleosynthesis)

$\sim 2\times10^9$ to $\sim 5\times10^8$ K

Formation of the First Protons, Neutrons (?)
(QCD Phase Transition)

$\sim 10^{13}$ to $\sim 10^{12}$ K

Quark-Gluon Plasma Era (?)

$\sim 10^{26}$ to $\sim 10^{13}$ K

Reheating

Inflation (??)

Grand Unified Era (???)

$\sim 10^{32}$ to $\sim 10^{28}$ K

Quantum Gravity Era (???)

$\sim 10^{32}$ K

13.8 billion yrs

9.8 billion yrs

9.2 billion yrs

~200 million yrs

380,000 yrs

~50,000 yrs

~1-20 min

$\sim 10^{-5}$ to $\sim 10^{-4}$ sec

$\sim 10^{-32}$ to $\sim 10^{-6}$ sec

$\sim 10^{-32}$ sec

$\sim 10^{-43}$ to $\sim 10^{-35}$ sec

$\sim 10^{-43}$ sec

$\sim 9\times10^{-30}$ g/cm^3

$\sim 1.3\times10^{-26}$ g/cm^3

$\sim 1.4\times10^{-26}$ g/cm^3

$\sim 10^{-26}$ g/cm^3

$\sim 3\times10^{-21}$ g/cm^3

$\sim 10^{-19}$ g/cm^3

~ 500 to ~ 1 g/cm^3

$\sim 10^{18}$ to $\sim 10^{14}$ g/cm^3

$\sim 10^{95}$ g/cm^3

$\sim 10^{95}$ to $\sim 10^{79}$ g/cm^3

$\sim 10^{71}$ to $\sim 10^{18}$ g/cm^3

Origin of Dark Matter

Matter-Antimatter Asymmetry Established

Conditions Studied by...

The First Cyclotrons

The Large Hadron Collider

At the Edge of Time

Nature shows us only the tail of the lion. But there is no doubt in my mind that the lion belongs with it even if he cannot reveal himself to the eye all at once.

—ALBERT EINSTEIN

SOME 13.8 BILLION YEARS AGO, our universe experienced its first moments. Almost everything about this earliest period of time remains a mystery to us. We don't know what forms of matter existed, or what physical laws they obeyed. All we know for sure is that our newborn universe had little in common with anything that exists in our world today.

Only a fraction of an instant later—10^{-43} seconds or so—gravity may have started to behave a lot like the force that we know and love. In contrast, the other known forces of nature—electromagnetism and the strong and weak nuclear forces—likely appeared very different from the way they do today.

Then things got really weird. Sometime in the first 10^{-32} seconds or so, our universe began to expand even more rapidly—wildly more rapidly—than it had been before. This expansion was so fast as to appear virtually instantaneous, and it left our universe utterly transformed. During this era, known as cosmic inflation, our universe grew in volume by an incredible factor of about 10^{75} within a span of only 10^{-32} seconds. This expansion carried every piece of matter away from every other at speeds far greater than the speed of light. By the time that inflation was over, every particle was left in isolation, surrounded by a vast expanse of empty space extending in every direction. And then—only a fraction of a fraction of an instant later—space was once again filled with matter and energy. Our universe got a new start and a second beginning.

After a trillionth of a second, all four of the known forces were in place, and behaving much as they do in our world today. And although the temperature and density of our universe were both dropping rapidly during this era, they remained mind-bogglingly high—all of space was at a temperature of 10^{15} degrees. Exotic particles like Higgs bosons and top quarks were as common as electrons and photons. Every last corner of space teemed with a dense plasma of quarks and gluons, alongside many other forms of matter and energy.

After expanding for another millionth of a second, our universe had cooled down enough to enable quarks and gluons to bind together, forming the first protons and neutrons. Within a few minutes, many of those protons and neutrons had fused together into the first atomic nuclei. During this era, our entire universe resembled the core of a modern star. But things did not remain that way for long. As space further expanded, the temperature continued to fall, leading to dramatic changes from minute to minute, and then from hour to hour and day to day. After a few hundred thousand years, our universe had cooled to a mere few thousand degrees. It was around this time that electrons began to bind to nuclei, forming the first complete atoms.

Slowly but steadily, clumps of matter began to collapse under the influence of gravity, which in turn led to the formation of stars and galaxies. We estimate that the first stars appeared in our universe about 200 million years after the Big Bang. These early stars were much larger and shorter lived than those found today. Only now are telescopes powerful enough to produce images of these stars about to come into use. Our Sun and its Solar System are relative latecomers to our universe, forming about 9.2 billion years after the Big Bang. Today, at the age of 13.8 billion years, our universe continues to expand, cool, and evolve. Its past contained a vast and diverse range of eras and transformations. Its far future is sure to usher in new epochs, both expected and otherwise.

If you've ever watched a documentary, listened to a lecture, or read a book about the Big Bang, there is a good chance that you've been presented with a timeline similar to the one described over the past few pages. Over most of this cosmic history, we have good reasons to be confident that the events and eras it describes actually took place. We have directly observed the formation of stars and galaxies, and we have measured the light released during the formation of the first atoms with remarkable detail and precision. We have determined the rate at which our universe has expanded over the past several billion years and identified the abundances of the various nuclear elements that were forged in the heat of the Big Bang. Taken together, this body of empirical evidence makes it clear that our universe has, in fact, evolved very much as our calculations had predicted—at least over the majority of its history.

But as it turns out, cosmic timelines are ordered not only by the progress of events, but also in terms of our direct knowledge. The events that lie in the most distant past, and closest to the Big Bang, are also generally those that we know the least about.

For the period of time ranging from a few hundred thousand years after the Big Bang to the present, we have a rich array of observations and measurements on which to rely, and this collection of data leaves us confident that we understand this portion of our universe's history quite well. Along with the vast majority of other cosmologists, I would be shocked if this part of the chronology turned out to be substantively wrong—there are just too many different and powerful lines of evidence that support our current understanding of this series of events. Finding out that we got this very wrong would be like finding out that there had never been an American Civil War, or that Christopher Columbus actually landed in Wales in the twelfth century and not in the West Indies in 1492. While it's good to keep an open mind about what you might have gotten wrong, in some cases the evidence is just too strong to reasonably contemplate being entirely mistaken.

As we go back farther in cosmic history, however, our confidence begins to decline. Between the first few seconds and a few hundred thousand years after the Big Bang, we have fairly strong support for what's described in the standard timeline. Observations and measurements tell us that the rate of expansion and the quantities of matter and energy in our universe cannot have been very different from those our calculations predicted. That said, it is still plausible that important and unknown cosmological events may have taken place during this period. The information we have about our universe's first hundreds of thousands of years is significant, but it is not exhaustive.

But reaching even farther back in time—into the first seconds and fractions of a second after the Big Bang—we transition from having incomplete information to having essentially no direct observations on which we can confidently rely. This era remains hidden from our view, buried beneath as-yet-impenetrable layers of energy, distance, and time. Our understanding of this period of cosmic history is, in many respects, little more than an informed guess, based on inference and extrapolation. Yet it is clear that these first moments are the key to many of our most urgent and enduring cosmic mysteries. Understanding this era is essential to understanding our universe.

With this book, I offer you a glimpse of the Big Bang—our universe's first seconds and fractions of a second. During this earliest of epochs, matter and energy took on very different forms from those found in our universe today and may have been subject to forces that we are yet to discover. Key events or transitions that we don't yet know about may very well have taken place. Matter likely interacted in ways that it no longer does, and space and time themselves may have behaved differently than they do in the world that we know. Almost everything that we know about physics could have been different during this first instant of time.

By any reasonable standard, the science of cosmology has had a spectacular century. One hundred years ago, we knew nothing about our universe's distant past and certainly nothing about its origin. But building upon Einstein's vision of space and time, astronomers discovered that our universe is expanding, and by the late 1960s it was clear that it had emerged over billions of years from the hot, dense state we call the Big Bang. For the first time, human beings had begun to understand how their universe began.

Since then, cosmologists have steadily pieced together the history of our universe, from these first moments to the present.

The past several decades have witnessed a diverse array of high-precision measurements that have enabled us to reconstruct our universe's past in new ways and with unprecedented detail. By measuring the expansion rate of our universe, the patterns of light released in the formation of the first atoms, the distributions in space of galaxies and galaxy clusters, and the abundances of various chemical species, we were able to confirm that our universe had expanded and evolved in just the way that the Big Bang theory had long predicted. Our universe looked more comprehensible than ever before.

And yet, not all is understood. Despite our considerable efforts, there remain essential facets of our universe that we simply do not know how to explain—especially pertaining to the first seconds and fractions of a second that followed the Big Bang. When it comes to the origin and youth of our universe, mysteries continue to abound.

The most famous, perhaps, is that of dark matter. Astronomers and cosmologists have determined how much matter there is in our universe to a very high degree of precision, and it is much more than exists in the form of atoms. After decades of measurement and debate, we are now confident that most of the matter in our universe does not consist of atoms or of any other known substances, but of something else that does not appreciably radiate, reflect, or absorb light. Over the past few decades, physicists have been engaged in an ambitious experimental program seeking to reveal what this substance is and how it was formed in the Big Bang. But despite initial optimism, we remain ignorant of dark matter and its nature. The experiments have performed just as designed, but have seen nothing. Dark matter has turned out to be far more elusive than we had once imagined.

Even the origin of "ordinary" matter harbors stubborn secrets of its own. Although protons, neutrons, and electrons, and
the atoms they constitute, can be easily created through well-
understood processes, such processes also create an equal
quantity of more exotic particles, known as antimatter. Whenever particles of matter and antimatter are brought into contact
with one another, both are annihilated. So why, then, does our
universe contain so much matter and so little antimatter? In
fact, why is there any matter at all? If matter and antimatter had
been created in equal amounts in the heat of the Big Bang—as
our current understanding of physics would lead us to expect—
then almost all of it would have been destroyed long ago, leaving our universe essentially devoid of atoms. Yet there are atoms
all around us. Somehow, more matter than antimatter must
have been created in the first fraction of a second of our universe's history. We don't know how or when this came to pass,
or what mechanism was responsible. But somehow, something
about the conditions of the early universe made it possible for
the seeds of atoms—and all of chemistry, including life—to
survive the heat of the Big Bang.

Going back even farther in time, we come to what is perhaps
the single most intriguing of our cosmic mysteries. In order to
make sense of our universe as we observe it, cosmologists have
been forced to conclude that space, during its earliest moments,
must have undergone a brief period of hyperfast expansion. Although this epoch of inflation lasted only a little longer than a
millionth of a billionth of a billionth of a billionth of a second,
it left our universe utterly transformed. In many ways, one can
think of the end of inflation as the true beginning of the universe that we live in. Despite identifying many compelling reasons to think that inflation really took place, cosmologists still

know and understand very little about this early, key era of our cosmic history.

In the 1990s, cosmologists set out on an ambitious program to measure the more recent expansion history of our universe, allowing us to determine the geometry and ultimate fate of our world. For the first time, it was thought, we would be able to learn whether our universe will continue to expand forever, or instead eventually reverse and collapse in upon itself. These measurements were ultimately successful, but they revealed to us something that very few scientists expected: our universe is not only expanding, but is expanding at an accelerating rate. To explain this fact, we have been forced to conclude that our universe contains vast amounts of what is known as dark energy, filling all of space and driving it apart. But our best efforts to understand this phenomenon have come up almost entirely empty-handed. We simply do not understand what dark energy is, or why it exists in our universe.

Each of these puzzles and problems is deeply connected to the first moments that followed the Big Bang. Whatever the dark matter consists of, it was almost certainly formed in the first fraction of a second after the Big Bang. Similarly, the simple fact that atoms exist in our world reveals that the earliest moments of our universe's history must have included events and interactions that we still know nothing about. Cosmic inflation also took place during these earliest of times, and its possible connection to the existence of dark energy raises many questions of its own. In these and other ways, our universe's greatest mysteries are firmly tied to its first moments.

In recent years, scientists have constructed new observatories and carried out new experiments in the hope of pulling back the shroud that has up until now hidden our universe's first moments from our view. But despite having conducted a range of impressive observations and measurements, we are in many ways only more perplexed than we were two decades ago. Instead of neatly resolving what had once seemed like a few loose ends, our increasingly precise cosmological measurements have only intensified the aforementioned puzzles, and even brought some new ones to light. As of late, it seems that the more we study our universe, the less we understand it.

Perhaps more than any other experimental or observational program, particle physicists and cosmologists alike had put a great deal of hope and confidence in particle accelerators such as the famed Large Hadron Collider in Geneva (LHC for short). These incredible machines accelerate beams of particles— typically protons or electrons—to the highest speeds possible, then collide these beams into one another. When protons collide at the LHC, many different kinds of matter can be created, including all of the known particle species, from electrons and photons to Higgs bosons and top quarks. In the early universe, interactions of these kinds filled all of space with a zoo of subatomic particles, all of which were constantly interacting with each other and being repeatedly created and destroyed. By studying these processes at the LHC, we have learned not only how matter and energy behave in our world today, but also how it behaved a minute, a second, a millionth of a second, and even a trillionth of a second after the Big Bang.

Many of us imagined that the new and spectacular LHC would lead us to a qualitatively superior understanding of our universe and its origin, enabling us to resolve many of our most puzzling questions. But since the start of its operations in 2010, the results from the LHC have in many ways left us only more confused. With the exception of the Higgs boson, this machine has not yet discovered any of the new particles or other phenomena that we anticipated. The problems cosmologists faced prior to the LHC remain firmly in place. Many of the solutions we had once imagined would resolve these problems are, in fact, not solutions at all.

For example, by remaining stubbornly elusive, dark matter has become only more perplexing as the experiments carried out over the past years and decades have ruled out many of our most promising hypotheses for what this substance—or substances—might consist of. In light of these results, cosmologists have been forced to give up their favorite theories and to consider radically new ideas about what the dark matter might be and how it might have formed in the first instants after the Big Bang.

It is from this perspective that I sometimes find myself contemplating the state of cosmology. We have in our possession a beautiful and remarkably successful theory. But it is also the case that we have recently struggled, if not outright failed, to explain many of our universe's most striking features. From the origin of atoms and the mystery of cosmic inflation, to the natures of dark matter and dark energy, it is clear that we are missing key elements in the way we understand our universe and its beginning.

We are at a time of reckoning. When it comes to understanding our universe and its origin, incredible progress has been made—there is no question about that. But despite this progress, it is undeniable that we are facing many formidable questions and vexing problems. Perhaps these issues are just a series of loose ends, which we will nicely tie up in the years ahead with new experiments and observations. But more and more often these days, I find myself wondering whether these problems might represent more than loose ends. Perhaps they are the symptoms of a deeper problem with the lens through which we see our world.

With this book, I will bring you on a tour of our universe's first moments. We will begin with Einstein's revolutionary insights into the nature of space and time and see how these ideas led to the discovery that our universe had a beginning in the hot, dense state we now call the Big Bang. I will do my best to explain how we learned what we know about our universe's early history, using tools ranging from telescopes to particle accelerators. From there, we will turn our attention to the puzzles, mysteries, and open questions that litter our universe's first fractions of a second. How did our universe come to contain so much matter and so little antimatter? How did the dark matter come to be formed? Our universe seems to have undergone a brief period of hyperfast expansion, but how and why? And is this connected to the fact that our universe is now once again expanding at an accelerating rate?

If you are looking for a story with an ending that wraps up nicely, you may have chosen the wrong book. This book is as much about open and unanswered questions as it is about anything we currently understand. But today's mystery is tomorrow's discovery. Powered by new data, observations, and ideas, we are poised to shed light on many of our most perplexing questions. With these new advances, we will see deeper and more clearly into the past than ever before—closer to the edge of time.

A World of Time and Space

The most beautiful thing we can experience is the mysterious. It is
the source of all true art and science. He to whom this emotion is
a stranger, who can no longer pause to wonder and stand rapt in
awe, is as good as dead: his eyes are closed.

—ALBERT EINSTEIN

HOW DID OUR UNIVERSE BEGIN? What was it like when it
was young? How has it changed and evolved over time? How
will it change in the future? These are the questions at the heart
of cosmology.

Considering how much we know about our universe and its
history today, it is remarkable that only a little over a century
ago, there was nothing one could call a science of cosmology.
In the first decades of the twentieth century, if you wanted to
ask a question about our universe's origin or its distant past, the
only people offering answers were theologians. Science knew
nothing—and claimed to know nothing—about these topics.
Many at the time thought that science would never be able to
address these kinds of questions. No reasonable person would
make that argument today.

Over the course of the past century, the work of countless physicists and astronomers has enabled us to develop a detailed understanding of our universe's 13.8-billion-year history, from the extremely hot and dense state known as the Big Bang through the formation of the subatomic particles, nuclei, and atoms that constitute our world today. Modern cosmologists understand how and why galaxies, stars, and planets formed. And with a high degree of confidence, we can describe what our universe was like as early as a few seconds after the Big Bang. Unlike people from any other time in history, we know what we are looking at when we look upon the night sky. And although there remains much to be learned, we know a great deal about our universe's past and how it came to be the way that it is today. Throughout the vast majority of human history, it must have seemed inconceivable that we could ever know such things. Today, anyone in the world with an internet connection has access to an almost unlimited collection of such facts.

What made it possible for us to put an end to our ignorance of our universe's origin, and begin to build a reliable scientific understanding of its nature and physical history? In part, this comes down to advances in technology. The telescopes that existed more than a century ago were simply not capable of capturing many of the things that would later make observational cosmology possible. Other key tools, such as particle accelerators, had not yet been invented. But there was also something else—something more basic—that had prevented us from seeking and uncovering our universe's cosmic secrets. The fact is, at the beginning of the twentieth century, physicists didn't yet understand enough about the fundamental natures of energy, matter, space, and time to even speculate about how our universe might change or evolve or how it might have begun. Before we could begin to answer—or even ask—questions about the origin of our universe, we needed a new and more powerful foundation to build upon. We were waiting for Albert Einstein's theory of relativity.

For over 200 years, our understanding of the physical world was founded upon the work and ideas of Isaac Newton. Over the time between the publication of Newton's masterwork, the *Principia*, in 1687 and the beginning of the twentieth century, physicists made advances in a huge variety of subjects including heat, electricity, magnetism, and light. The ways in which scientists thought about these phenomena, however, remained fundamentally confined to a Newtonian perspective of the world. Newton's ideas about motion, force, and momentum were successfully applied to problem after problem. There seemed to be no limit to what these physical principles could be used to explain. New generations of scientists continued to solve new puzzles while maintaining the same underlying vision of our universe that had been established so long ago by Newton.

The philosopher Thomas Kuhn famously argued that science does not progress in steady, incremental steps, but instead through dramatic transformations called paradigm shifts. In these upheavals, it's not just new facts that are introduced into a scientific community. Instead, an entirely new worldview takes hold, leaving in ruins the old ways of thinking about a subject. To those whose minds are entrenched within the old worldview, a paradigm shift can be difficult to grasp and may even seem nonsensical. In the years leading up to the 1920s, for example, physicists were actively debating whether light was a wave or was instead made up of a collection of particles—two possibilities that each lie well within the boundaries of a Newtonian worldview. But these scientists couldn't imagine that light might take the form of particles—photons—that were also each individually waves. Within the existing paradigm, a wave consisted of a collection of objects. The possibility that a single object could be both a particle and a wave led to many strange and counterintuitive conclusions and was inconceivable before the paradigm shift we now call quantum mechanics. The answer to the question of whether light is made up of particles or waves turns out to be none of the above, and both of the above, at the same time. From the vantage point of the old thinking, a new paradigm can seem ridiculous and utterly absurd. It reminds me of when Bob Dylan sings, "The sun's not yellow, it's chicken." To make sense of it, you have to forget something about what you think the words mean, and only then begin to build up a new way of understanding the ideas involved.

Relativity's rise to prominence was an outright and utter paradigm shift, and one of the most significant in the history of science. Before Einstein, physicists were capable of thinking about their world essentially only in Newtonian terms. And while this long-standing worldview was undeniably powerful, it also had real limitations. For one thing, it provided no way to address questions of cosmology. From the Newtonian perspective, it made no sense to ask how the size or shape of space might change with time, or how space or time might have begun. Cosmology simply had no place within the system put forth by Isaac Newton.

With the introduction of relativity, Einstein tore down and destroyed much of what physicists thought they had known. When the dust settled, Newtonian physics had been replaced with something very different: an altogether new and beautiful way of envisioning our world and its physical laws. It is not an overstatement to say that almost everything we know about space and time can be attributed to Albert Einstein.

*   *   *

To us, living in our world, space and time are at the heart of existence. We conceive of objects with spatial extent—height, length, width—and at locations in space. Events, from our perspective, take place at moments in time. Things change only by moving through space as time passes, or by transforming from one thing into another at a given moment in time. Time and space are the essence of our world, and they form the very basis of our imagination.

As infants, the cognitive ability to recognize geometrical patterns—lines, edges, shapes—is among our first neurological developments. As a consequence of our biological hardwiring, human beings tend to find themselves unable even to imagine a reality that isn't based upon these concepts. Existence and being seem to rely implicitly upon the existence of space and time. Without these constructs, our imagination becomes limited and impotent. To our human minds, for something to exist is for it to exist within space and time.

It is perhaps not surprising that physics has always had the concepts of space and time at its foundation. From the viewpoints of Aristotle, Galileo, or Newton, the laws of physics are, when it comes down to it, simply rules for determining how the locations of objects within space change with time. Laws of physics are laws of motion. Without space and time, we can't talk about what it means for something to be moving, or for something to be close to or far from something else. We can't talk about something happening without a notion of time for it to be happening in. Even the concept of energy is built upon space and time, because energy is ultimately nothing more than motion or the potential for motion. Without time and space, nothing can change, and without change, it is hard to imagine any reality worth imagining. Without time, nothing happens. Without space, nothing is.

In the course of our daily lives, space is relatively simple—high school geometry covers everything we need to understand it. Time might be a bit more subtle, but we still perceive it in a relatively straightforward way. Over the past century, however, physicists have come to appreciate that space and time are not so simple or straightforward. Unlike Aristotle, Galileo, and Newton, we now understand that space and time can change and evolve, and be shaped, stretched, and deformed. Space and time can expand, contract, wrap, twist, disconnect, inflate, or even begin or cease to exist. To Newton or Galileo, it would have been unthinkable that such a range of verbs could be applied to space or time. But it is precisely these dynamic and animated characteristics of space and time that the science of cosmology is built upon.

Imagine yourself walking upon a vast and featureless surface. You come across Newton's car with its keys in the ignition. (You can tell that it's Newton's car by the apple-shaped hood ornament.) You get in and drive until the odometer reads one mile, then turn 90 degrees to the right, drive another mile, and so forth, until your route covers a complete square. When you get out of the car, you notice on the ground your own footprints from when you first walked up to the car. Just as you had expected, you are back to exactly where you started.

Newton's car behaves just as our intuition tells us it should. In particular, it moves through what mathematicians and physicists call Euclidean space. Euclidean space, named for the ancient Greek philosopher Euclid, obeys fives basic rules, sometimes called axioms or postulates. These postulates consist of statements that seem perfectly uncontroversial: any two points in space can be connected by a straight line, for example; all right angles are equal to each other; and that for any straight line, there is exactly one straight line parallel to it that passes through any given point in space. This last postulate implies, among other things, that parallel lines never cross one another. There's a good chance that in high school geometry you were taught each of these postulates as indisputable facts. They seem so obvious, it can be hard to imagine that any of them could be wrong. From this perspective, Newton's car behaves in the only way that it could.

Until the nineteenth century, Euclid's postulates were universally considered to be self-evident and indisputable. The German philosopher Immanuel Kant—best known for his work in epistemology, or inquiries into how we can determine whether something is true—went as far as to argue that even without an opportunity to study or observe the world, we could have come up with our concepts of space and time using only pure thought and reason. In other words, he believed that Euclid's vision of geometry was the only logical possibility, or at least the only way in which human beings could conceptualize space.

By the first half of the nineteenth century, however, mathematicians had begun to stretch their imaginations beyond the confines of Euclid's system. In particular, a number of mathematicians—including Janos Bolyai, Nikolai Ivanovich Lobachevsky, and Bernhard Riemann—had managed to develop self-consistent geometrical frameworks that did not adhere to Euclid's fifth postulate—the one pertaining to parallel lines. In these new "hyperbolic" and "elliptic" geometries, two parallel lines need not remain parallel. Instead, two straight lines that are parallel to each other at one point in space can come together or diverge from one another as you follow them along their paths. In these geometrical systems, it can be shown that the sum of the three angles of a triangle can be greater than or less than 180 degrees and that the ratio of a circle's circumference to its diameter need not be the number π. Within these non-Euclidean systems, a great deal of what you learned in high school geometry turns out not to be true.

But just because a mathematician can write down a weird geometrical system does not mean it is real in any physical sense. Sure, mathematics has proven very useful in helping us to understand our physical world, but not all mathematical possibilities are necessarily realized in nature. We may be able to rationally imagine a world with some strange rules of geometry, but that does not mean that those strange rules are followed in our world. Mathematicians had managed to prove that Euclid's fifth postulate did not, according to the dictates of logic alone, have to be true. Whether or not it was actually true in our physical world remained an open question.

Intrigued by these strange new systems of geometry, a handful of mathematicians and physicists began to consider whether they might have anything to do with the physical world. But despite a few intermittent shows of interest, most physicists did not take these exotic geometries seriously—until, that is, Albert Einstein placed them at the heart of his general theory of relativity.

Once again imagine yourself on that vast and featureless surface, walking up to a car. But this time, there is no apple-shaped ornament on the hood. This is not Newton's car, and something tells you that driving this car is going to be a very different experience. As you sit down behind the wheel, you notice a strange-looking man in the passenger seat. You worry for a moment that his long, messy hair might catch fire from the pipe that he is smoking. In a heavy German accent, he says to you, "Welcome to my car. Let's go for a drive."

Just as you had done before in Newton's car, you begin to drive around the perimeter of a square, each time turning when your odometer tells you that you have traveled a mile. The strange man sitting next to you makes you increasingly nervous, however, so you find yourself driving faster and faster as the journey progresses. When you finish driving the four miles, you step out of the car and notice, to your surprise, that you are not exactly where you started. You think for a minute that the car's odometer must be broken, but the strange man assures you that his car is in perfect working order. "There is nothing wrong with my car," he says with confidence. "Perhaps the world through which you are driving is not as simple as you had imagined."

Things become even weirder whenever you drive Einstein's car near an object that contains a large amount of mass or other energy—such as a star or a planet. In the presence of such energy, you find yourself drifting inexorably toward the object, despite trying to drive in a straight line. Somehow, the very shape of space is distorted.

To understand why this is happening, think back for a moment to Euclid's fifth postulate: for any straight line there is exactly one straight line parallel to it that passes through any given point in space. Within the geometrical framework of Einstein's masterpiece—the general theory of relativity—this axiom is simply not true. And without this axiom to rely on, the concept of a straight line becomes very different from what your intuition might lead you to expect. It is indeed strange to think that a straight line could follow an arc through space, but there is more than one way to define a straight line. One way is to say that a line is straight if it follows the shortest route between two points in space. In Euclidean space, the shortest route between any two points is an ordinary, intuitive straight line. But when mass or energy deforms the shape of space in a region, the shortest route between those points might not seem straight to us at all. What Einstein taught us is that straight lines through space can be curved by the presence of mass or energy.

According to Newton's laws of motion, in the absence of any outside influences, an object will continue to move at the same speed and in the same direction, along a straight line. In Einstein's view, this is still true, but Einstein's idea of a straight line is not the same as what Newton had in mind. When mass or other energy distorts the shape of the surrounding space, the straight lines within that space become curved. When you drive Einstein's car near a space-distorting object, your trajectory curves in the direction of the object, just as if you were being pulled toward it. This distortion of space and time acts just like an attractive force—just like gravity. In fact, this is exactly what gravity really is.

With this connection, Einstein showed us that gravity is not merely a force. It is instead a manifestation of the geometry of space and time. The presence of mass and other forms of energy change the very shape of our world, curving and distorting it. These distortions cause objects to move in a way that is almost identical to how they had long been predicted to move under the influence of Newtonian gravity. What Einstein described as the curvature or distortion of space and time by mass and energy is, in fact, the very phenomenon that for hundreds of years had been known simply as gravity.

In 1915 Einstein completed his masterpiece—the general theory of relativity. By explaining the phenomenon of gravity not only as a force, but as a consequence of geometry, Einstein over-turned hundreds of years of established physics. And although many physicists admire this theory for its profound beauty and mathematical elegance, its greatest virtue is that it is true. By this, I mean that its predictions agree extremely well with observations. To date, no experiment or test has ever been found to conflict with the predictions of general relativity. Perhaps one day we will discover a circumstance under which Einstein's theory fails, but nothing like that has happened so far.

Under most conditions, Einstein's theory predicts that objects should move in almost exactly the same way predicted by Newtonian gravity. This is of course good news, since Newton's theory predicts so many things so well. There are exceptions to this, however. For example, Newtonian gravity predicts that Mercury's orbit should be slightly different from what astronomers actually observe. And although Newton gets this prediction slightly wrong, general relativity gets it right. Einstein's theory also correctly predicts how light should be deflected as it passes by massive bodies—something that was observationally confirmed for the first time during a solar eclipse in 1919. More recently, scientists have seen the effects of general relativity in numerous high-precision measurements. Even the Global Positioning System (GPS) would not work if it did not take into account the effects of general relativity. In order to do their job, GPS satellites have to keep time within an accuracy of about 20 nanoseconds. But according to general relativity, time passes differently for the satellites than it does on the surface of the Earth, because of the differences in the Earth's gravity and the corresponding curvature of space. Without taking general relativity into account, this system would struggle to determine locations within a kilometer, much less the few-meter precision that we have become accustomed to.

In order to use Einstein's theory to make predictions such as these, one has to solve what are known as Einstein's field equations. And although these equations are exceedingly difficult to manipulate, they are conceptually fairly simple. At their heart, they relate two things: the distribution of energy in space, and the geometry of space and time. From either one of these two things, you can—at least in principle—work out what the other has to be.

So, from the way that mass and other energy is distributed in space, one can use Einstein's equations to determine the geometry of that space. And from that geometry, we can calculate how objects will move through it. When space is flat—without curvature—objects move along what we intuitively think of as straight lines. But near large quantities of energy, space and time are curved, and objects move through space along arcs or other non-straight paths. The Earth doesn't move along an ellipse around the Sun because of an attractive force of gravity. Instead, energy in the form of the Sun's mass has warped the geometry of space throughout the Solar System, and the Earth is simply moving along the most direct route possible, which happens to be an orbit in the form of an ellipse. The very presence of this energy has distorted the space and time that surrounds it. From this perspective, gravity isn't really a force at all. Instead, it's the direct consequence of the geometry of space and time.

When Einstein published his general theory of relativity in 1915, he didn't seem to have had cosmology in mind. As far as we know, he hadn't imagined that his theory was going to tell us anything about the history of our universe, its future, or its origin. But the equations of this theory can be used to predict how space and time will respond to the presence of mass and other energy. And this means that if one knows the contents of our universe, they can use these equations to determine the geometry of our universe as well, and to predict how it will evolve and change with time.

Within only a couple of years' time, it became apparent to Einstein and others that his theory was capable of explaining more than how and why objects move through our universe. It also provided us with a new and powerful way to understand our universe itself.

A World without a Beginning?

Never in all the history of science has there been a period when
new theories and hypotheses arose, flourished, and were abandoned
in so quick succession as in the last fifteen or twenty years.

—WILLEM DE SITTER (1931)

BY THE LATER MONTHS OF 1916, Einstein had started to think
about a new application for his theory of general relativity. Although the ink on his theory was still wet, he was no longer
content to apply these equations to the orbits of planets or to
the deflection of light. Instead, he had started to think about the
shape and structure of our universe as a whole. In a letter to a
colleague and friend, Einstein joked that this new work "exposes
me to the danger of being confined to a madhouse." (At least
I think he was joking.) This wasn't the kind of topic that physicists worked on in 1916. Even for Einstein, this was a bold and
unorthodox line of inquiry.

If you want to use Einstein's field equations to calculate the
path of a planet's orbit around a star, you start with the masses,
locations, and velocities of the astronomical bodies in question.

If you're good at solving sets of nonlinear differential equations, you can use this information to work out how the surrounding space and time is curved, from which you can then determine how things move. Whether you were calculating the trajectory of a fly ball at Wrigley Field or of a spaceship falling into a black hole, the process would be essentially the same. But these are not the only kinds of problems that Einstein's theory can address. In principle, you can start with any distribution of mass and energy and use Einstein's equations to determine the consequences. You can even use these equations to calculate the overall geometry and evolution of our universe.

When Einstein carried out his first cosmological calculations, he assumed that matter was distributed evenly throughout the entire universe. This was a simplifying assumption, of course, intended to make the problem more tractable. In reality, the actual density of our universe is not always homogenous: the density of the Earth, for example, is obviously higher (more than a trillion-trillion times higher, in fact) than the average value in outer space. On larger scales, however, it turns out that our universe is fairly uniform. For example, if I were to count the number of galaxies in two different billion-light-year-wide regions of space, I would find a very similar number of galaxies in each. So, even though he couldn't have possibly known it at the time, Einstein's approximation turned out to be a good one.

Once Einstein had specified—or at least correctly guessed—how matter is distributed in the universe, he set out to use his equations to deduce the large-scale geometry of space and time. The answer to this question depends on how much matter our universe contains. If the overall density is high enough, all of space throughout the universe will be positively curved, and the geometry will be what mathematicians call elliptic. If you

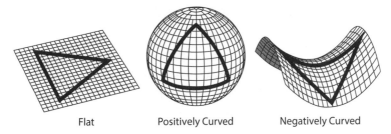

Flat   Positively Curved   Negatively Curved

Examples of two-dimensional surfaces that are geometrically flat (*left*), positively curved (*center*), and negatively curved (*right*). In flat space, the rules of ordinary Euclidean geometry apply: parallel lines remain parallel and the three angles of any triangle add up to 180 degrees. In spaces that are either positively or negatively curved, these rules that we were taught in high school geometry simply don't apply.

follow two parallel lines far enough through such a universe, they will eventually converge, and the angles of triangles will always add up to more than 180 degrees. If the density is lower, space will instead be negatively curved—a hyperbolic geometry. In this case, lines that are parallel at one point recede from each other as you follow them through space, and the angles of triangles always add up to less than 180 degrees.

Although pondering the geometry of our universe is interesting in its own right, Einstein's cosmological investigations revealed something else that was stranger still. Contrary to his expectations, the equations seemed to insist that our universe's geometry should be changing with time. Almost everything in our universe is in a state of motion. And Einstein's equations imply that space itself should be in motion as well—either expanding or contracting. The one thing that it cannot do is stay the same.

This fact carries with it enormous implications. Prior to Einstein, space was seen as entirely unchanging and eternal. But if space is capable of change, then it may have had a beginning, and perhaps may have an end. Without even intending to,

Einstein had forced us to consider the cosmic history and origin of our universe.

When Einstein looked at the cosmological consequences of his theory, he didn't like what he saw. For some reason, Einstein was deeply uncomfortable with the idea of an expanding or contracting universe. This was always difficult for me to understand. Einstein was a profoundly creative human being, and no one has ever accused him of suffering from a lack of imagination. But for some reason, he carried a strong philosophical prejudice that our universe must be eternal and unchanging. To him, it just seemed obvious—something like common sense—that our universe was always the way that it is now, and that it would remain this way forever. But about this, Einstein was simply wrong.

Let's step back for a minute and consider more carefully what physicists mean when they say that the geometry of a universe is changing. As we discussed in the previous chapter, we tend to imagine space as a fixed background through which things move, but general relativity taught us that space isn't actually like that. The presence of energy—including mass—can warp and curve the surrounding space and time, altering the way in which objects move through it. Similarly, the distance between any two points in space doesn't have to be the same at different points in time. When we say that space is expanding or contracting, what we really mean is that the distances between fixed objects in space are increasing or decreasing. Similarly, the volume of a region can change as time passes, making the size of that space larger or smaller than it was in the past.

According to Einstein's equations, a universe with a high density of matter will not only be positively curved, but will also ultimately contract, bringing all points in space closer together as time progresses. On the other hand, a lower density universe will have a negatively curved geometry and will expand forever.

At the time, astronomers knew little about the large-scale structure of our universe. They knew that our Sun was one of billions of stars that made up the Milky Way, but it still wasn't clear whether the Milky Way was a unique system, or if instead there were other galaxies like it elsewhere in our universe. Furthermore, the Milky Way didn't appear to be in the act of expanding or contracting. The stars in the night sky did not seem to be moving toward or flying apart from one another en masse, and our Solar System had remained largely unchanged for billions of years. Nothing astronomers had observed thus far suggested that our universe was evolving. To them, with the technology available at the time, our universe appeared to be static. All of this must have helped to reinforce Einstein's prejudice.

Although he of course understood his equations and was aware of what they were telling him about our universe, Einstein nevertheless continued to insist that our universe could not be changing. He was determined to find a static solution to his field equations. The trouble was, as originally written, these equations simply don't have a static solution—space must either expand or contract. To find the kind of solution he was looking for, Einstein had no choice but to change the field equations themselves.

As it turns out, however, it is not so easy to change the field equations of general relativity. Unless you resort to some very radical propositions—such as introducing the existence of spatial dimensions beyond the three we know and experience—there is really only one self-consistent way to change the structure of these equations. This change consists of adding a new piece to the field equations—something that Einstein called the cosmological term.[1] This term acts like a force that pushes outward and drives the universe to expand. If assigned precisely the right value, this part of the equation could balance against our universe's tendency to contract, preventing it from collapsing in on itself.

Einstein knew that this cosmological term was an entirely ad hoc addition to the field equations; unlike the other components of his theory, there was nothing that we knew about gravity that justified it. But its inclusion was also perfectly consistent with the mathematics of the theory, and there were no observations that told us that it couldn't exist. In this sense, Einstein was perfectly within his rights to include the new cosmological term. And it did seem to make a static universe possible. At the time, that was good enough for Einstein.

In 1917 Einstein published the first paper to consider the cosmological consequences of his general theory of relativity. In it, he described a universe that he argued followed directly from his field equations. This hypothetical universe soon became known among physicists as "Einstein's World."

So what was Einstein's World like? First of all, aided by the cosmological term, it was static: it did not expand or contract with time. The shape and size of Einstein's World were the same in the past as they would be in the future. According to Einstein's cosmology, our universe did not evolve, and it had no origin.

---

1. The cosmological term is proportional to the cosmological constant, which you might have heard of elsewhere. For the purposes of this book, you can treat these two phrases as interchangeable.

Second, the geometry of Einstein's World wasn't flat or Euclidean, but instead was positively curved. And perhaps most intriguing of all, the volume of his universe was finite, but without having any edges or boundaries. In Einstein's World, space wraps around on itself. In other words, if you were to move far enough in any one direction, you would eventually come back to where you started.

To understand this concept, it helps to consider an analogy. Although space in Einstein's World is three-dimensional, it has a number of properties in common with the two-dimensional surface of a sphere.[2] For example, if you move westward along the Earth's equator, you will eventually circumnavigate the globe and return to where you started. Similarly, in Einstein's World, if you were to travel in a given direction for a few hundred million light years—or about 10^{22} miles—you could circumnavigate the entire universe. And much like Einstein's World has a finite volume, the Earth's surface area is finite—about 200 million square miles—and without any edges or boundaries.

To my eyes, Einstein's World contains a great deal of beauty. It is essentially an unchanging and perfectly uniform example of what is known as a three-sphere—the three-dimensional analog of an ordinary sphere's two-dimensional surface. It's easy to see why Einstein was drawn to it. But like so many beautiful things, Einstein's World turned out to be an illusion. It may contain truth in some kind of mathematical sense, but not in a

2. Thinking about the surface of the Earth as its own two-dimensional space—as opposed to part of the ordinary three-dimensional space—can sometimes be confusing. In my experience, it helps to remember that I am talking in this analogy only about the surface of the Earth—there is no up or down in this two-dimensional space, only North–South and East–West.

physical one. The universe we live in is very different from the one Einstein originally envisioned.

Although Einstein was the first to consider the cosmological consequences of the general theory of relativity, others quickly followed. Within the span of only a few years, several physicists had discovered their own cosmological solutions to the field equations. Einstein's theory, they quickly learned, allows for many different and sometimes strange kinds of universes.

From among the cosmological solutions identified during this period of time, one stands out as the most important—or, at least, the most like our universe. This solution was discovered in 1922 by a young Russian physicist named Alexander Friedmann.

Friedmann's greatest scientific accomplishments might have come even earlier if not for the intervening events of history. Shortly after he finished graduate school, the First World War broke out, and the newly founded Imperial Russian Air Service put him to work as a bomber pilot on the Austrian front. Although he survived the war, tranquility remained elusive for Friedmann. His new home—the city of Perm, near the Ural Mountains—was a constant battlefield, with both the Communist and anti-Communist armies fighting to take and retake control of the city. In 1920, Friedmann moved once again, this time to Saint Petersburg (recently renamed Petrograd), where he held a number of academic posts. It was when Friedmann was there that the civil war finally came to an end, allowing him a chance to return to his science and begin to contemplate the implications of Einstein's new theory.

In many ways, Friedmann's cosmological investigations began from a similar starting point as Einstein's. Both he and Einstein assumed a universe that was uniform in density, and like

Einstein, Friedmann included a cosmological term in his calculations. But Friedmann saw no reason to insist that the geometry of our universe must be fixed and unchanging with time; in fact, in stark contrast with Einstein, he became convinced that our universe *must* either expand or contract. Regardless of Einstein's own philosophical objections, Friedmann was beginning to show that the theory of general relativity does not allow for the possibility of an unchanging universe. Change is simply inevitable.

When Einstein first heard about Friedmann's work, he was almost immediately convinced that it must be wrong—he even thought that he had identified a mistake in the calculations. Over the course of a year or so, however, Einstein began to back down, eventually conceding that Friedmann's math appeared to be correct. But even then, Einstein continued to insist that the expanding and contracting solutions Friedmann had discovered weren't physically possible. While he acknowledged that Friedmann's results were "both correct and clarifying," Einstein maintained that "a physical significance can hardly be ascribed to them." The universe, according to Einstein, could not be changing.

With the benefit of hindsight, we can see that Einstein was simply wrong about this. For one thing, it can be shown that Einstein's cosmological solution is unstable, and doesn't describe a truly static universe. The static nature of Einstein's World relies critically on the assumption of a perfectly uniform distribution of matter. In this sense, Einstein's World is like a pencil, perfectly balanced on its tip. In theory, the pencil might be able to stay upright forever, but in practice, a stray air molecule could tip it over at any time. In the same way, if our universe were like the one proposed by Einstein, then those regions with the most matter would slowly start to collapse, while those with the least matter would start to expand. Because our universe is not perfectly—but only approximately—homogeneous, it can't be static. But at the time, this conclusion was not yet widely appreciated. To most scientists, it was not at all clear who was right.

Throughout this chapter, I've been making references to both Einstein's field equations and their solutions. But I haven't really explained the distinction between these two things, what they are, or how they are connected. In any physics textbook, you'll find equations that govern how things move and change, but these equations alone don't tell the whole story. The same set of equations can sometimes have many different solutions, allowing for many different possibilities, any of which might exist in the world.

Consider something as simple as the trajectory of a baseball.[3] Although Newton's equations—describing the force of gravity and its effect on motion—are useful for predicting this kind of thing, I can't tell you where a given baseball is going to be at a given time from the equations alone. First of all, I would need to know the characteristics of the environment: assuming it's moving near the surface of the Earth, I would need to plug the Earth's mass and radius into the equations in order to determine the strength of gravity acting on the ball. Second, in order to predict where the baseball will be at some time in the future, I would need to specify its location, velocity, and spin at some other point in time—what physicists call the initial conditions of the problem. Under most circumstances, a physicist can

---

3. For those of you who aren't baseball fans, feel free to substitute a cricket ball or whatever other projectile you prefer.

calculate the motion of something only if they have all three of these things at hand: the appropriate equations, the relevant features of the environment, and the initial conditions.

The same holds true when considering the origin and evolution of our universe. Early cosmologists, including Einstein, Friedmann, and others, all had the appropriate equations for the problem—the field equations of general relativity. But they were more or less speculating about the environmental factors, as well as the initial conditions. Each proposed solution had a different value for the current rate of expansion or contraction, the average density of matter, and the cosmological term. It was as if these early cosmologists were trying to guess how the baseball would move from the equations alone, without knowing where it had started from or even what planet the game was being played on.

That being said, there are some conclusions regarding how a universe will evolve over time that can be drawn from Einstein's equations alone. As I've said before, any universe without a cosmological term will either expand or contract—there are no static solutions. Furthermore, any such expanding universe will always be slowing down, while any contracting one will always shrink more quickly with time. This is like noticing that all baseballs follow arcs—parabolas—as they travel through space. Although you might not be able to say which arc a given baseball will follow, you know that it will follow some arc all the same.

By adding the cosmological term to his equations, Einstein had enabled general relativity to take on a series of other solutions. In some of these solutions, the universe expands faster and faster—like a baseball accelerating without limit, flying off into deep space. In other solutions, the universe can remain balanced—at least for a time—between the forces of expansion and contraction, like a baseball that stops moving entirely and simply levitates without change.

To determine the trajectory of a baseball, you need more than the equations that govern its motion—you need observations as well. The same is true for the evolution and origin of our universe. Throughout the 1920s, physicists continued to argue the merits of their favorite cosmological solutions, with no consensus in sight. The science of cosmology may have started from theoretical considerations, but these could take us only so far. Fortunately, the era of observational cosmology was about to begin.

Throughout the history of science, theoretical arguments and mathematical reasoning have settled relatively few debates. Much more often, it's new observations and data that end arguments and change minds. This was certainly the case regarding the disagreement between Einstein, Friedmann, and others over whether our universe is static, expanding, or contracting. The math had been litigated and relitigated, and yet there was no sign of any emerging consensus. This debate could not be resolved by arguments of pencil and paper alone. Instead, it required a new generation of powerful telescopes.

In the first decades of the twentieth century, a majority of astronomers thought of our universe as the Milky Way—as far as they were concerned, these were interchangeable terms. Although today we recognize that the giant collection of stars and gas that we call the Milky Way is merely one galaxy of very many within the larger universe, our galaxy was at the time the only one known to exist. It was as if astronomers were living on an island, without any way of knowing whether there were other islands in the sea. For all they knew, our galaxy could have been one of a countless number of such concentrations of matter. Or perhaps the Milky Way was instead a truly unique and solitary collection of stars, surrounded in all directions by a vast and endless ocean of empty space.

In reality, astronomers had observed quite a few galaxies by this time—they just hadn't realized what they were looking at. The challenge was that a distant galaxy looks a lot like a much smaller and relatively nearby cloud of gas. Astronomers observed many of these nebulae, but couldn't tell from their images whether they were galaxies or not. To figure out what these objects really were, astronomers needed a new way to measure their distances. Fortunately, in 1908, the American astronomer Henrietta Leavitt discovered just such a technique. Leavitt's method used a special class of pulsating stars, known as Cepheids, for which there turns out to be a reliable relationship between how often they emit pulses of light and how bright their pulses are. This relationship enabled astronomers to determine the true brightness of a given Cepheid, as well as the distance to it. For the first generation of observational cosmologists, this was a game-changing advance.

During this same period, telescopes themselves were becoming much more powerful. The bigger a telescope is, the more light it can collect, which in turn enables its user to observe fainter and more distant objects. In 1919, the biggest telescope in the world was the brand-new Hooker Telescope, located at the Mount Wilson Observatory outside of Los Angeles.

Using the Hooker Telescope and Leavitt's technique, Edwin Hubble was able to accurately measure the distances to many Cepheid stars, including several that resided within various nebulae. Many astronomers had long thought that these nebulae were part of the Milky Way, but Hubble's measurements revealed that some of them are much too remote to be part of our own galaxy. For example, he found that the Cepheid stars located within the Andromeda Nebula were a whopping 900,000 light years away—well beyond the limits of the Milky Way.[4] In making this measurement, Hubble discovered that the Andromeda Nebula was not merely some nearby cloud of gas and stars, but a full-fledged galaxy, similar in size and shape to the Milky Way. It is for this reason that what had long been known as the Andromeda Nebula is now called the Andromeda Galaxy.

Within a few years, this debate ended once and for all. We now know that the Milky Way is an island of stars within a much larger universe, containing many, many other such islands. Modern estimates are that there are roughly a trillion galaxies within the potentially observable volume of our universe. In a sense, this discovery represented a new advance in the Copernican Revolution. The Earth is just one of the planets in orbit around the Sun, which is merely one of the billions of stars in our galaxy. Hubble took this yet another step forward. Our own galaxy—the Milky Way—is not unique or particularly special, but is instead only one of many galaxies within a much greater universe.

Hubble's discovery that Andromeda and many other nebulae are themselves galaxies—not unlike the Milky Way—carried enormous implications. But in itself, this new information didn't reveal anything to us about whether or not our universe is expanding or contracting. To determine whether or how our universe might be evolving, astronomers needed not only to measure the distances to a set of objects—such as galaxies— but also how fast those objects are moving. If our universe is

---

4. Hubble's value for this distance turned out to be an underestimate. The true distance to Andromeda is approximately 2.5 million light years.

contracting, then the galaxies being studied by Hubble should all be moving toward us. And if our universe is expanding, then they should all be moving away, being carried away by the stretching of space itself.

Fortunately, the theory of general relativity provides astronomers with a way to determine how fast a given galaxy is moving. Much like the pitch of a sound wave shifts when its source is moving toward or away from you (the Doppler effect), Einstein's theory predicts that the frequency of light will be shifted if its source is in motion. The astronomer Vesto Slipher was among the first to recognize this application of Einstein's new theory and carried out measurements of this kind for many galaxies. More often than not, Slipher found that the light from these galaxies was shifted toward lower frequencies, or redshifted—indicating that they were moving away from us.

Over the years to follow, Hubble and his colleague Milton Humason continued to use the state-of-the-art Hooker Telescope to observe and study galaxies, building up an impressive catalog of distance measurements for forty-six of these objects. When they combined their distance measurements with Slipher's velocities, they saw a clear pattern beginning to emerge. In particular, the most distant galaxies they had observed—each several million light years away—were all moving away from us at considerable velocities. It was as if we were located in the middle of a giant cosmic explosion, with everything flying away from us in all directions. But even more interesting was that the velocities of these galaxies were roughly proportional to their distance from us—a relationship that has since become known as Hubble's Law. While the most distant galaxies in Hubble's sample are zipping away from us at speeds of around 1,000 kilometers per second, the most nearby galaxies are moving much more slowly. For every million light years or so that lie between us and a given galaxy, that galaxy will be moving away from us at a speed of about 21.7 kilometers per second, or about 50,000 miles per hour.[5]

Initially, Hubble and Humason weren't sure how to interpret this result. But what they had in fact discovered is that our universe is not static, regardless of Einstein's insistence to the contrary. Furthermore, the data showed that we are not at the center of some cosmic explosion. While intriguing, that hypothesis cannot explain the observed proportionality between galaxies' distances and their velocities. Instead, the very space that constitutes our world is expanding with the passing of time. Distant galaxies appear to be moving away from us because the amount of space between them and us is growing.

If you're thinking about the concepts of expanding or contracting space right now for the first time—and maybe even if you're not—these ideas may seem a little difficult to get your head around. If you're like most people, there is a good chance that you are imagining it incorrectly, at least in some respects. But don't feel bad. Just about everyone gets it wrong at first. I know I used to. It is a very strange idea, and we have very little natural intuition for it. To better understand what it means for space to be expanding, let's consider an analogy.

Imagine that you are standing on the surface of the Earth and that the Earth is expanding—like an inflating balloon. As this balloon-like Earth expands, the distance between any two points on the Earth's surface grows. For example, the distance between Chicago and Detroit is 240 miles, as the crow flies. But

5. I'm using the modern values for these numbers here. Hubble's original measurements were plagued with systematic errors that led him to underestimate the distances to galaxies by a factor of about seven.

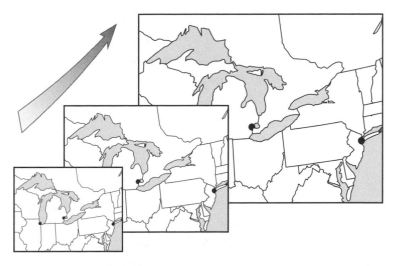

If the Earth were expanding like an inflating balloon, the distance along its surface between any two cities would be increasing with time. Every city in this example is moving away from every other at a speed proportional to their distance from each other. This is the same relationship that Edwin Hubble observed for galaxies.

if the Earth gradually expands to twice its original radius over a period of an hour, the Chicago–Detroit distance will increase from 240 to 480 miles over that hour. So if you happen to be in Chicago, Detroit would appear to be moving away from you at a speed of 240 miles per hour.

Now if you were to look from Chicago toward any other city, you would find that city to be moving away, although not necessarily at the same speed. New York City, for example, is about three times farther away from Chicago than Detroit is, so New York will be moving away from Chicago about three times more rapidly—about 720 miles per hour. So as the Earth expands, every location on the Earth's surface moves away from every other at a speed proportional to their distance from each other. This is precisely the same relationship that Hubble had found for galaxies.

The same analogy can help us to understand some other aspects of expanding space as well. From any location on the expanding Earth's surface, all cities are moving away from you. And just as there is no center of the surface of the Earth, there is no center of the expanding universe. Any observer, located anywhere in our universe, will observe the same recession of galaxies that Hubble discovered.

When I'm explaining this idea in a classroom or in a public lecture, it's usually around this time that someone asks, "But what is space expanding *into?*" Most people picture expanding space as a process of space growing into, or gradually taking up, some other region of space—like the *volume* of an inflating balloon. But this misses the point of what we mean when we say "space." Space can't expand into other space. When we say that space is expanding, we mean all of space, not just some of it. There is nothing for space to grow or expand into. If there were, we would call that thing *space.* The space of our universe is getting larger, but without moving into anything else.

When cosmologists talk about space expanding, they aren't talking about galaxies or other objects moving into some pre-existing, previously unoccupied space. It isn't that galaxies are flying apart, flowing into the volume of a larger universe. Instead, space itself is becoming larger as time passes. All points in space are farther apart from one another today than they were yesterday.

When thinking about the physics of everyday objects—things like flying baseballs or braking cars—we benefit a great deal from our intuition. Millions of years of natural selection have made our brains quite adept at picturing and understanding these kinds of phenomena. But not all ideas in physics—and especially modern physics—are so intuitive. Expanding space is a perfect example. Without sufficient effort and training, our brains can struggle to conceptualize this. But the human brain is also flexible and can be coaxed into understanding some very strange ideas.

If you find yourself struggling with the concept of expanding space, I have a mental trick to suggest: instead of space expanding, imagine a universe in which all objects are shrinking. To understand what I mean, consider any two points in space. If you measure the distance between them and find that it is increasing with time, you would conclude that this is because the amount of space between them is increasing—space is expanding. But how do you actually go about measuring a distance between two points in space? You need some kind of measuring standard—the equivalent of a ruler or a yardstick. At this very moment, I am sitting in a room of my local coffee shop. If I wanted to determine the width of the room, I could take out a yardstick—or failing that, use the span of my arms or something else along those lines—and count how many lengths it takes to reach from one wall to the other. Without a measuring standard, this would be impossible. If I gave you a picture of a featureless room without anything in it—basically an empty cube—and asked you to estimate its size, you would have no way of doing so. It could be a centimeter across, or a mile. Without a standard to compare to, you can't determine a distance. Distances are meaningful only in comparison to others.

So distances in space are dependent on what we use to measure them. As a consequence, we have at least two very different ways that we can choose to think about expanding space. The first, and more conventional, of these is to think of the amount of space in our universe as increasing. But alternatively, we could instead imagine that everything within space is shrinking, while space itself remains unchanged. Both ways of thinking about this are completely equivalent and indistinguishable from each other. If the room that I am sitting in is not changing, but my yardstick is steadily shrinking, then the room will appear to be larger every time I measure its width. It will appear to be expanding. But in order for me to be convinced that the room is really growing, not only would my yardstick have to be shrinking, but everything else in the room as well, including my own body. If absolutely everything in the room were shrinking at the same rate, there would be no way of telling whether the room was growing or the stuff in it was shrinking.

Similarly, one can feel free to think of our universe as expanding or, instead, of everything in our universe as shrinking. When I say everything, however, I mean *everything*. For example, we often use the time that it takes for light to travel between two points in space as a measurement of the distance between them. So, in order for all of our measuring standards to be shrinking, even the speed of light would have to be slowing down with time. Similarly, the distances between protons, neutrons, and electrons within atoms would have to be shrinking, as well as the distance between the Sun and Earth and, well . . . everything. As you can imagine, it would be very challenging to find an explanation for why all of these properties of nature would be changing at the same rate, making it rather implausible that this could be mimicking the expansion of space. But that being said, if you find the idea of space expanding without expanding into anything to be confusing, you might find it helpful to instead imagine a static universe in which everything is shrinking.

Hubble's observations not only revealed that our universe is expanding, but also began to tell us about our universe's past, as well as its future. If you know the velocity of a baseball at any one moment in time, you can use Newton's equations to predict how it will move in the moments ahead and to deduce where it had been in the moments prior. In much the same way, Hubble's measurement of our universe's expansion rate, combined with the equations describing Friedmann's cosmological solution, made it possible for scientists to begin to deduce the very history of our universe.

Running Friedmann's equations backward in time, we learn that our universe was more compact in the past and contained much higher densities of matter and energy than it does today. In this compressed state, our young universe was once very hot—and hotter still as you go back farther in time. In fact, if you extrapolate these equations far enough—billions of years into the past—they describe a period of time in which every last corner of our universe was filled with an ultra-hot soup of energetic particles.

Despite Einstein's efforts to show that our universe could somehow be static, the collective work of Friedmann, Hubble, and other early cosmologists showed that this is not the case. According to the equations of general relativity, the size and shape of space simply must evolve as time advances. Furthermore, the observation that our universe is expanding suggested that our entire universe—everything we know and everything we will ever experience—grew and expanded out of a hot primordial state, over billions of years, into the world that we see around us today. Decades later, this primordial state would become known as the Big Bang. For the first time in history, humanity caught a true glimpse of the origin of all things.

4

Glimpses of the Big Bang

Telescopes are in some ways like time machines. They reveal
galaxies so far away that their light has taken billions of years to
reach us. We in astronomy have an advantage in studying the
universe, in that we can actually see the past.

—MARTIN REES

I CAN STILL vividly remember my first encounters with cosmology. It was my senior year of college, and I had just begun
to learn about the subject. I found myself absolutely awed by
the incredible events that played out as our universe expanded
and cooled. Almost as amazing as the science itself was the fact
that human beings were able to discover and understand this
deep history of our universe—not by speculation or myth-
making, but by observation, measurement, and reason.

The basic elements of this cosmic history—collectively called
the Big Bang Theory—are extremely well founded in observational evidence. No serious scientist today doubts that our universe expanded from a very hot, dense state over the past 13.8
billion years, forming nuclei, atoms, galaxies, stars, and planets
along the way. The evidence in support of this chronology is

simply too overwhelming to deny. But this was not always the case. When these ideas were first proposed, few scientists embraced this view of our universe's history, and many actively opposed it. Most physicists and astronomers began to view this theory more favorably only after a considerable body of evidence had accumulated in its favor. This is a key feature of any healthy scientific community—its members, although often skeptical, are eventually willing to change their minds when the evidence calls for it. And the evidence in favor of the Big Bang has called loudly, howling from the rooftops.

Since the 1990s, observational cosmologists have collected an astoundingly detailed and precise body of data, entirely unlike anything that had previously been possible. This data not only confirmed the essential elements of the Big Bang Theory, but also revealed an increasingly vivid picture of many of the events that took place over the course of our universe's history. Thirty years ago, few scientists would have believed that we would learn so much about our universe so quickly. But nonetheless, we certainly have. It is an exciting time to be a cosmologist.

Hubble's discovery that galaxies are moving away from us with velocities proportional to their distances provided the first clear evidence that our universe is expanding and evolving over time. This fact carries with it incredible implications. After all, if something—such as our universe—can change, then that thing cannot be truly permanent or eternal. And something transient may, or even must, have had a beginning. It was inevitable that scientists would begin to ask questions about the very birth of our universe.

Among the first generation of cosmologists, most were not focused on our universe's distant past or origin. A clear exception to this was Georges Lemaître. In a world in which science and religion are so often opposed to one another, Lemaître is a refreshing figure. He was not only a mathematician and astronomer, but also a Catholic priest. And although religion has often worked to hinder scientific progress, it is perhaps not surprising that it was a religious man who first considered as a scientific hypothesis the possibility that our universe and everything in it was created in a primordial event.

Lemaître was introduced to the fledgling science of cosmology in 1923 as a graduate student at Cambridge, where he worked with the famous astronomer and physicist Arthur Eddington. After stints at both Harvard and the Massachusetts Institute of Technology, he returned to his home country of Belgium, where he was a part-time lecturer at the Catholic University of Leuven. At this point, little about Lemaître stood out. Although he had earned degrees from some very impressive institutions, his original research had made little impact. But that was all about to change.

Lemaître's first major work on cosmology was published in 1927. Among other calculations, he used the equations of general relativity to deduce Hubble's Law, relating the velocity and distance to a given galaxy—two full years before Hubble made his famous observations. According to Lemaître, this relationship was a necessary consequence of general relativity. But many were skeptical, including Einstein himself. While Einstein acknowledged to Lemaître that "your calculations are correct," he also quickly added that "your grasp of physics is abominable."

It is probably the case that if Lemaître's career had ended at this point, few would remember him now. But in the wake of Hubble's discovery, Lemaître prepared to publish another and even bolder idea. This work would set the stage for the next century of cosmology.

In 1931, Lemaître published a short but profound paper titled "The Beginning of the World from the Point of View of Quantum Theory." In only a few paragraphs he laid out a proposal in which everything in existence—every particle of matter and every photon of light—arose from the decay of a singular "atom." Despite its name, Lemaître's primordial atom had little in common with the entities on the periodic table. For one thing, it had a truly gargantuan mass, equal to the total energy of everything in the universe today.

Before this solitary primordial atom decayed, Lemaître argued, the very concepts of space and time simply had no meaning. After all, it is possible to measure distances only *between* objects, and if there is only one object in the universe, then there are no distances to measure. Similarly, in this solitary state, Lemaître's atom could do nothing: there was effectively no space to move in, and nothing to interact with. In a universe with only one object, there can be no space. In a universe in which nothing happens, there can be no time. In this sense, space and time came into existence with the decay of the primordial atom.

What Lemaître described has, of course, little in common with the Big Bang as we now understand it. But this paper presented the first physical and scientific proposal for the origin of our universe—a creation event from which all other events would follow. At the time, many scientists had begun to accept that our universe is expanding, but the idea that space and time themselves might have had a beginning was seen as almost universally objectionable—"repugnant," in the words of Arthur Eddington. A notable exception to this was Einstein, who eventually came to hold a much more favorable view of Lemaître and his physics. After attending a lecture of Lemaître's in 1933, he commented, "This is the most beautiful and satisfactory explanation of creation to which I have ever listened."

*   *   *

As time passes, the expansion of space steadily dilutes the matter and energy contained within it. Thus, in the past, our universe contained a higher density of matter than it does today, and in the future, the density of our universe will become lower still. As matter and energy are diluted, they also cool, so our universe was hotter in the past and will be cooler in the future.

This relatively simple logic is the key to the Big Bang Theory. If you follow it backward through time, you can begin to reconstruct the history of our universe. About 3.1 billion years ago, our universe was approximately half of its current size, by volume. Going back even farther, to 8.5 billion years ago, it occupied only a tenth of its current volume. And 12 billion years ago, it was only a hundredth as large as it is today.

In these earlier eras, everything was closer together, and thus denser. But it was more than just the overall density that was different in our universe's distant past. The relative quantities of matter and other forms of energy also changed as space expanded. In the case of matter, density evolves just as our intuition leads us to expect. As the volume of space increases by a factor of 1,000, for example, the density of matter in that space goes down by a factor of 1,000, so long as no matter is destroyed or created in the process. But things are not so simple for photons or other particles traveling at or close to the speed of light. In addition to the ordinary dilution that all particles experience—the aforementioned factor of 1,000—individual photons are stretched by the expansion of space, which increases their wavelength and reduces their energy and temperature. As a result of this effect, known as the cosmological redshift, the energy in light is diluted more rapidly than ordinary particles of matter. As the volume of space increases by a factor of 1,000, the total energy contained in photons falls by a factor of 10,000—ten times more than that experienced by matter.

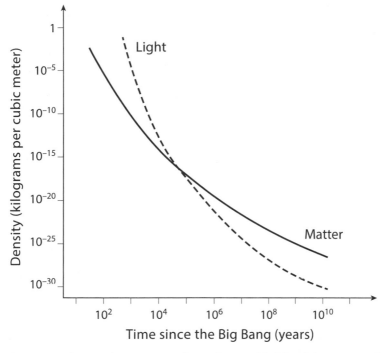

Over cosmic history, the expansion of space has steadily diluted the amount of matter and light contained within our universe. In addition, the energy in the form of light has been further reduced as a result of cosmological redshift. Thus in the distant past, the ratio of light to matter was much higher than it is today.

As a consequence, whatever the ratio of light to matter is in our universe now, we know that it must have been higher in the past. And although our universe contains much more energy in the form of matter than in light today, if you look back far enough in time, you will find an era in which this was not the case. Our universe began not in a state of matter, but in a state of light.

Throughout the first half of the twentieth century or so, no one was sure exactly where the atoms in our universe came from or how or when they were formed. But if you had to place a bet on the question, the good money seemed to be on the possibility that their nuclei originated in stars.

Since the early 1920s, physicists had argued that the fusion of hydrogen into helium nuclei could be capable of generating most of the energy that stars release. Stars consist mostly of hydrogen and, according to $E = mc^2$, even a small fraction could be transformed into enough energy to power a typical star for billions of years. By 1940 or so, most of the basic nuclear processes at work within stars had been identified and understood. In the most massive stars, these processes could form atomic nuclei as heavy as iron. Even heavier elements—such as gold, silver, and uranium—could be synthesized in the explosions of stars, called supernovae. To many scientists at the time, it looked as though every nucleus in our universe, and across the entire periodic table, may have had its origin in a stellar furnace.

But by the late 1950s, this theory, known as stellar nucleosynthesis, had begun to show some strain. In particular, it was becoming increasingly clear that stars alone could not be responsible for the large quantities of helium found throughout our universe. Helium represents a huge fraction of the atoms in our universe—about 25 percent by mass. There seemed to be no way that so much helium could have formed in stars.

Despite this challenge, stellar nucleosynthesis seemed at the time to be the only plausible way the elements could have formed. After all, nuclear fusion occurs only at temperatures of billions of degrees or higher, and such temperatures are found only in the cores of stars. But in 1946, the Soviet-American physicist George Gamow—once a student of Alexander Friedmann—suggested another possibility. Realizing that the observed expansion of space seemed to imply that our universe was much hotter in the distant past, Gamow proposed that most of the nuclei may have formed then. Instead of within stars, nuclei may have been forged in the heat of the Big Bang.

A few years ago, I went back and read Gamow's original 1946 paper and the follow-up paper he wrote in 1948 with his student Ralph Alpher.[1] I can't think of any other papers in the history of science that are both so important and so wrong. The proposal that atomic nuclei may have been formed in the Big Bang is of profound significance, and the last thing that I want to do is to diminish that. But that being said, there were numerous problems with Gamow and Alpher's calculations. For example, Gamow and Alpher didn't correctly take into account the way that nuclei repel each other, and thus overestimated many of the rates for nuclear fusion. And even more importantly, they did not seem to appreciate that only the lightest few nuclear species could be produced in any significant quantities in the Big Bang—deuterium, helium, and lithium. The early universe had been hot for too short a time to produce the heavier elements.

Within a few years, a number of different physicists had identified and corrected these problems. When the dust settled, it appeared possible that the bulk of the hydrogen and helium in our universe could have been generated in the Big Bang. Most of the heavier elements, however, could not have formed in this way and so must have originated in stars. For this reason, stellar nucleosynthesis remained the predominant theory for some time. We now know that nuclei actually formed through a combination of these two mechanisms. While most of the nuclear

---

1. As part of a brilliant marketing ploy (and because Gamow thought it was funny), the physicist Hans Bethe was listed as an author of this second paper, despite the fact that he did not contribute to the research or to the actual writing. Instead, his name was included only so that the author list would read "Alpher, Bethe, Gamow," leading it to become known as the "alpha-beta-gamma" paper. This paper is still widely known by this name today.

species across the periodic table formed in stars, the lightest few elements were overwhelmingly generated in the first few minutes of our universe's history.

During the primordial era in which the first nuclei were being formed, our universe was filled with intense heat and light. The quantities of helium and other nuclear species that were synthesized in the Big Bang depend critically on how many photons were present during that time. By the 1950s, it was clear that if the observed helium did have its origin in the Big Bang, then there must have been an era—roughly the first 100,000 years after the Big Bang—in which light, and not matter, constituted most of the energy in our universe. Although the energy carried by these photons was gradually reduced as our universe expanded through the process of cosmological redshift, these primordial photons did not disappear from our universe. In fact, these particles of light have survived to the present day, existing now as a sea of invisible microwaves that fill the entirety of our universe.

This cosmic radiation—a remnant from our universe's distant past—is everywhere and permeates all of space. The temperature of this radiation—known as the cosmic microwave background—is currently a frigid 2.728 degrees above absolute zero, or about 455 degrees below zero Fahrenheit. It was not always so cold, however. When this radiation originated about 13.8 billion years ago, and a mere 380,000 years after the Big Bang, it was a blistering 3,000 degrees—roughly the temperature of the surface of a star. In contrast with the cold and largely empty state that we find our universe in today, the extreme heat of this radiation once occupied the entirety of our young universe. Nowhere was there emptiness. No corner of space was hidden or protected. All of space was filled with energy and heat.

*   *   *

As our universe expands, it steadily cools. But some changes in temperature have a greater significance than others. For instance, as liquid water cools, its characteristics change gradually—such as increasing modestly in density as it loses energy. But when water reaches 0 degrees Celsius, it transforms into ice. Something analogous to this transition happened when our universe was relatively young, about 380,000 years after the Big Bang. It was then that the temperature of our universe dropped below 3,000 degrees for the first time. And 3,000 degrees is not just any temperature—it's a key thermal landmark. It's the freezing point of atoms.

People don't usually think of individual atoms as something that can freeze, but they can. If you have ever taken a chemistry class, you probably recall how groups of atoms or molecules can freeze to form a solid, and how a solid can melt to become a liquid or a gas as the individual atoms or molecules become unbound from one another. This, however, is not the kind of melting and freezing that I am talking about here. I am not talking about groups of atoms interacting with each other; instead, I am describing something that can happen to an individual atom.

During the first few hundred thousand years after the Big Bang, all of space was filled with protons, electrons, and small groups of protons and neutrons bound together as nuclei. The pieces that make up atoms were present, but there were no atoms themselves. But as our universe cooled, the electrons began to bind themselves to protons and helium nuclei, forming the first complete atoms. Before this time, our universe had been simply too hot for atoms to survive. At temperatures above a few thousand degrees, electrons are not bound to their nuclei, and it is impossible for atoms to remain intact. In other words, it is above their melting point.

The formation of the first atoms was *the key event* in cosmic history as far as observational cosmologists are concerned. Before the electrons and protons had become bound together into atoms, all of space was filled with a plasma of electrically charged particles—a plasma that was almost entirely opaque to light. For this reason, our telescopes can observe only the light that originated after this key event. Using a telescope to look farther back in time is as difficult as using one to look deep beneath the surface of the Sun.

As the first atoms formed, the opaque plasma transformed into a gas of electrically neutral hydrogen and helium atoms, through which light can travel with relative ease. In fact, most of the photons that were present shortly after this transition have been traveling through space ever since, without interacting with or being absorbed by anything. When cosmologists study the cosmic microwave background today, they are observing these very photons that were set free during this transition. And although this light gradually cooled as our universe expanded, the photons that make up the cosmic microwave background are otherwise basically the same as they were 380,000 years after the Big Bang. This radiation is to a cosmologist what fossils are to a paleontologist. By studying its properties, we can learn an enormous amount about our universe's history and past.

Though it surrounds us constantly, the cold background of cosmic radiation can be challenging for scientists to detect and study. Unlike the types of light that we see with our own eyes, the photons that make up the cosmic background have far less energy and far longer wavelengths. Our most common types of telescopes—optical or even infrared—cannot detect the light that was left over from the Big Bang. As the techniques of radio astronomy improved in the 1950s and 1960s, however, it became possible for the first time to get a glimpse of this relic from our universe's hot youth. But much as the first fossilized dinosaur skeletons were not found by people who were looking for them, the discovery of the cosmic microwave background was also something of an accident.

The first radio telescope to identify the cosmic background radiation does not look much like what most people imagine when they think of a telescope. Instead of a long tube-like object with lenses and mirrors, the Holmdel Telescope—named for its location in Holmdel, New Jersey—was a strange-looking 50-foot aluminum antenna. It was designed not to study the cosmos, but to communicate with satellites. Two radio astronomers named Arno Penzias and Robert Wilson, however, turned the Holmdel Telescope to their own purposes in the 1960s, using it to search for radio waves from astronomical sources throughout the Milky Way. But while they were scanning the skies, their antenna kept picking up an unexpected and persistent hiss. No matter where they pointed their telescope, and no matter when they made their measurements, there was a static hum that just wouldn't go away.

As you have probably guessed, this irritating hum was no ordinary kind of noise. The faint buzz in their antenna was nothing less than the very photons that had been traveling unimpeded throughout the universe since the time of the first atoms' formation. Without even knowing it, Penzias and Wilson had seen into our universe's deep past. They had witnessed the very light of the Big Bang.

Only after speaking with some cosmologists at Princeton did Penzias and Wilson begin to appreciate the significance of their discovery. Of course not everyone was initially convinced that this signal had anything to do with the formation of atoms in the early universe. The Big Bang Theory was still a fairly controversial idea in the 1960s, and many scientists favored other cosmological models. Some opponents of the Big Bang Theory, for example, argued that Penzias and Wilson had actually measured the scattered light from the accumulation of billions of stars throughout the universe, and not the cosmic radiation from our universe's hot past. At the time, it was hard to know who was right. To most scientists, the origin of this cold background of radiation was far from clear.

This question was further complicated by the fact that the theoretical predictions associated with the cosmic background radiation were—to put it kindly—somewhat uncertain. Unbeknownst to Penzias and Wilson, earlier physicists had already realized that a uniform background of cold radiation, present throughout all of space, should have been left behind by the Big Bang. They couldn't agree, however, on the precise characteristics that this background should have. Theoretical physicists including George Gamow, Ralph Alpher, Robert Herman, and Robert Dicke had all made estimates of the cosmic background's temperature. Some of these were pretty good, while others, in hindsight, were not. Alpher and Herman's first estimates in 1948, for example, were pretty close—they calculated that the radiation should have a temperature of about 5 degrees above absolute zero. While this might not seem very accurate compared to the true value of 2.728 degrees, most of this discrepancy can be attributed to the fact that the age and expansion rate of our universe were not known very accurately at the time. Given the impact that these uncertainties had on their calculations, I consider their estimate of 5 degrees to be hitting the nail on the head. A couple of years later, however, the same scientists backtracked and changed their estimate to

28 degrees—not exactly what I would call an improvement. In contrast, Gamow started off with a rather poor estimate of 50 degrees, but improved over time—estimating a temperature of 7 degrees in 1953 and then 6 degrees in 1956. He seemed to be slowly closing in on roughly the right answer, but not everyone agreed. Even in the 1960s, the individuals and groups working on this problem did not seem to be converging. This did little to inspire confidence that the observed background radiation had anything to do with the Big Bang. After all, how do you know whether your telescope is in fact measuring the radiation from the Big Bang when you can't even agree about what the radiation from the Big Bang is supposed to look like?

Eventually, however, the confusion came to an end. As astronomers continued to measure the spectrum of this radiation in more detail, and as theoretical physicists slowly moved toward better agreement among their predictions, it became increasingly clear that the observed radiation was, in fact, the very cosmic background that was predicted by the Big Bang Theory. And at least as important was the fact that the measured properties of this radiation were *not* consistent with the predictions of other models. Over a decade or so, these improvements led to a consensus that Penzias and Wilson had, in fact, seen traces of the Big Bang. In 1978, they were awarded the Nobel Prize in Physics for their discovery. The Big Bang Theory was no longer seen as speculative or controversial, at least among scientists and the scientifically informed. It had matured into what can only be called a scientific fact.

The detection of the cosmic microwave background confirmed that our universe had once been in a hot and compact state, filled with a dense plasma of charged particles. Our world did indeed emerge from a Big Bang. This incredible discovery did not mark the end of our quest to understand the origin of our universe, however. It was only the beginning.

Over decades, cosmologists have built upon the work of Penzias and Wilson, steadily learning more about our universe—a great deal more. For example, by measuring and analyzing the subtle variations in the temperature of the cosmic microwave background across the sky, we have been able to determine the geometry of our universe with remarkable precision. We now know that it is flat, or at least very close to flat, with no significant positive or negative curvature. These measurements, combined with observations of the large-scale distribution of galaxies and galaxy clusters, have also enabled us to determine the average density of matter in our universe—a minuscule 3.3×10^{-24} grams per cubic meter—to better than 3 percent precision, and the amount of time that has passed since the Big Bang to an amazing precision of better than 0.2 percent. We know a great deal about how our universe expanded over time and how galaxies and stars came to form. The current era has indeed earned its name—the age of precision cosmology.

And yet, despite all of this progress, many important questions remain unanswered—in particular when it comes to our universe's earliest moments. We still don't know, for example, how the protons and neutrons that make up the atoms in our universe survived the heat of the Big Bang. From what we do know, it seems that these particles should have gone extinct long before a single atom had had the chance to form. But something that we don't yet understand must have prevented this from taking place.

Other modern cosmological observations have raised as many or more questions than they have answered. For example, scientists have used observations of the cosmic microwave background to determine the quantity of protons, neutrons, and other forms of matter in our universe. These determinations reveal that only about 16 percent of the total matter is made up of protons, neutrons, or any kinds of atoms or molecules—84 percent is made of something else entirely. We can clearly tell that this other matter is there because of its gravity; we see the evidence of its gravitational pull in numerous observations. Yet we simply do not know what it is or how it was formed in the early universe. For lack of a better name, we call this mysterious stuff dark matter. But naming something is very different from understanding it.

But even dark matter cannot account for all of the observations modern cosmologists have made. Measurement of the cosmic microwave background, combined with those of our universe's expansion rate, have presented entirely new mysteries. A few decades ago, there was an almost uniform consensus among cosmologists that the expansion rate of our universe should be slowing down over time. But since the late 1990s, astronomers have observed and reported precisely the opposite behavior: over the past few billion years, our universe has been expanding faster and faster. To explain these unexpected observations, cosmologists have considered some very radical possibilities, ranging from reintroducing Einstein's cosmological term to altering the very theory of general relativity itself. Much as with dark matter, we see the mystery, but not yet the solution. Furthermore, whatever it is that is driving the acceleration of our universe's expansion rate requires a vast amount of energy—more than is contained in all of the matter in our universe. We don't understand what this energy is or why it is there, but once again we have named it. We call it dark energy.

* * *

Over the past century, we've learned a great deal about our universe and its history. And with every year that passes, we learn even more. Yet I think that it is fair to say that the birth and very origin of our world remains almost a total mystery. But without mystery, there can be no discovery. Employing new tools and new ideas, we relentlessly strive to unravel these mysteries and begin to understand our universe's earliest moments.

For many thousands of years, human beings have wondered and asked questions about the distant past and the remote future of our world. The fact that we continue to ask such questions does not distinguish us from our ancient ancestors. What does make us different is that for the first time in history we, as a species, are capable of producing real and credible answers to these questions. We are among the first to be witness to the Big Bang.

The Universe and the Accelerator

The Large Hadron Collider (LHC) is the most powerful telescope
ever built.

—JOHN ELLIS

IMAGINE FOR A MOMENT that you live on a tropical island,
and that you and everyone you know has always been on that
island. On the very coldest of all days, the temperature drops to
55 or maybe 50 degrees Fahrenheit, but never lower. There is
liquid water in every direction as far as the eye can see. But you
have never experienced ice.

Now imagine that your scientists have determined that the
average temperature on the island has been slowly but steadily
increasing, and that this seems to have been going on for quite
some time. They estimate that a few thousand years ago your
island's average temperature was only 50 degrees, and that
tens of thousands of years before that it often dropped below
30 degrees—colder than anyone living on the island has ever
experienced. Your fellow islanders wonder what it might have

been like back in those frigid days of the distant past. But without any experiences of real cold to draw on, how can they know or find out?

To answer this question, the islanders might set their sails for distant places, hoping that they'll find somewhere with a colder climate that they can study for themselves. But alternatively, they might one day invent the technology of refrigeration, allowing them to create conditions similar to those found long ago on their island—all without leaving the comforts of home.

Imagine the potential for discovery that comes with such a machine. By turning up the power of their refrigerator, the islanders could observe nature under conditions they had never before witnessed, enabling them to find out about the distant past, and even recreate the conditions of prehistory. With such a machine they could learn about a time when the water surrounding their island was frozen into solid ice—a state that would otherwise have been unimaginable to them.

To understand what our universe was like in its earliest moments, we too rely on machines that recreate the physical conditions of the distant past. But these machines are not refrigerators—they are, in fact, quite the opposite. Using particle accelerators, we study how matter and other forms of energy behave at the highest of temperatures—those that were found throughout our universe as early as a trillionth of a second after the Big Bang. Since we cannot travel back in time or otherwise directly witness the first instants of our universe's history, we recreate a microcosm of the Big Bang here on Earth.

During the first moments that followed the Big Bang, there was nothing in our universe that you or I would be able to recognize—even atoms did not exist yet. Instead, all of space was filled with a dense soup of energetic quantum particles.

When we think about everyday objects—such as baseballs or marbles—we can usually trust our intuition to gauge how they might behave or what might happen when they come into contact with each other. But when it comes to electrons, photons, quarks, and other kinds of quantum particles, we find that our intuition often fails us—sometimes badly. Such particles can be spontaneously created and spontaneously destroyed. They can do things that baseballs and marbles will never do.

Although the behavior of quantum particles can seem strange, their nature doesn't have to be a mystery. Like everything in our universe, quantum particles follow physical laws—rules that, with the right tools, we can learn and understand. Particle accelerators are the primary tools that physicists use to discover these rules.

The world's preeminent particle accelerator is the modern marvel known as the Large Hadron Collider—the LHC, for short. When describing the LHC, it can be hard to find vocabulary that adequately conveys the awe that many of my colleagues and I feel toward it. I want to tell you that the LHC is large, but it's not just large. It's beyond enormous, gigantic, or immense. Simply put, it is the largest and most complex machine that humankind has ever built. The LHC propels a beam of protons through an underground 27-kilometer circular tunnel, or ring, extending beneath the suburbs of Geneva, Switzerland, and into nearby France. As these particles travel around the tunnel, they are accelerated by a collection of over 1,600 powerful superconducting magnets. But these magnets are not just powerful; they are the most powerful magnets ever built. Most of them individually weigh tens of tons and are capable of generating magnetic fields more than 100,000 times as strong as the Earth's. In order for these magnets to function, they have to be maintained at an operating temperature of only 1.9 degrees above absolute zero, which requires a constant supply of nearly 100 tons of liquid helium. This makes the LHC the largest cryogenic environment in the world. In fact, we know of no region in the universe that is both so large and so cold. Even the vacuum of outer space is warmer in comparison.

Once the LHC's magnets have fully accelerated a beam of protons traveling around its ring, these particles reach a speed of 99.999999 percent of the speed of light. This is the highest speed to which human beings have ever accelerated anything. At this rate, a given proton goes all the way around the LHC's 27-kilometer ring about 11,000 times every second. These protons travel through the tunnel in bunches, each of which contains over 100 billion protons. At four designated locations around the ring, the proton beams are directed head on into each other, maximizing the total amount of energy that is present in each proton-proton collision. Every time that two protons collide at the LHC, a tremendous 13,000 gigaelectron-volts, or GeV, of energy is concentrated into one place at one time.

Surrounding each of the LHC's four designated collision points is a detector designed to measure and record many of the particles that are created in the LHC's collisions. Each of these detectors is nothing short of an engineering marvel. The ATLAS detector, for example, is a 46-meter-long and 7,000-ton conglomeration of twenty-first century materials and electronics. From a single collision, a detector such as ATLAS can record the energy and momentum of dozens or even hundreds of photons, electrons, muons, and other particles as they travel outward from the collision point.

Although only a small fraction of the protons that pass through a collision point within a detector actually strikes another proton, such collisions still take place at an astonishing rate of about 700 million times each second. If the LHC's detectors

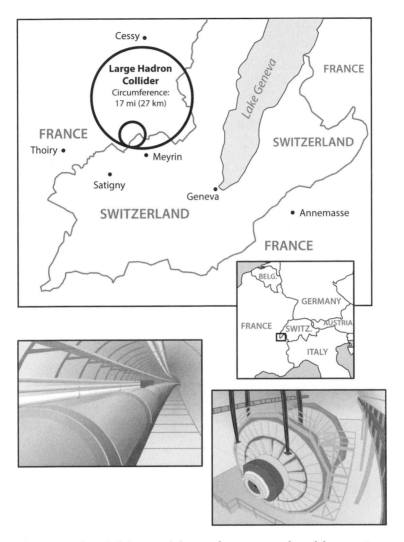

The Large Hadron Collider propels beams of protons around a 27-kilometer circular tunnel, accelerating them to nearly the speed of light. By colliding these protons together, physicists can study the kinds of interactions that took place only a trillionth of a second after the Big Bang.

were to record and keep all of the data associated with every collision, this would generate approximately 600 terabytes of data per second, far more than can practically be recorded with existing technology. To deal with this problem, the LHC's detectors use sophisticated algorithms to select—in real time—the most rare and otherwise interesting collisions, and to then retain and store this important subset of the total data. When all is said and done, about 25 gigabytes of data are permanently stored during each second that the LHC is operating, adding up to tens or even hundreds of petabytes each year. This data is then stored across a network of hundreds of computing facilities, distributed across more than thirty different countries.

An army of thousands of particle physicists—including many graduate students and other young scientists—descend on this data once recorded, sifting through it in the hope that it will enable us to understand more than we did before about our universe and its physical laws. In the most dramatic of moments, this culminates in the discovery of entirely new and previously unknown forms of matter—such as the Higgs boson, which was discovered in 2012. Discoveries such as this are the reason we build machines like the LHC. And the operation of this incredible machine is far from over. It is possible that the Higgs will turn out to be only the first of the particles that are discovered at the LHC in the years and decades to come.

But perhaps equally important to the discovery of new particles are the precision tests that machines such as the LHC provide of our existing theories. Over the past several decades, the theory known as the Standard Model has dominated the field of particle physics. This theory describes the seventeen known fundamental particles and their interactions and provides us with a detailed set of predictions for how each of them should behave and interact. So far, the predictions of this theory have been amazingly successful. Take the magnetic moment of the electron, for example. This quantity describes how an electron will respond to a magnetic field, and its measured value matches the prediction of the Standard Model to within one part in a trillion, making it the most accurately predicted quantity in the history of science. This level of accuracy is like successfully predicting the diameter of the Earth to within one hundredth of a millimeter or the mass of the Eiffel Tower to within a few milligrams.

Physicists at the LHC have compared the predictions of the Standard Model with their data in thousands of different ways, applying an enormous battery of tests. By precisely measuring the ways in which each of the known particles is created, decays, and otherwise interacts, we can test the Standard Model and see if its predictions are right. So far, they seem to be. In every respect that we have been able to test, the Standard Model appears to provide a very good description of our universe at energies as high as thousands of GeV. This also means that the Standard Model provides us with a good description of how our universe behaved when it was only a trillionth of a second old and at a temperature of nearly 10^{17} degrees.

When the LHC's proton beams are directed into one another, the collisions that occur are very similar to those that occurred repeatedly among particles in the early universe. By studying these kinds of high-energy interactions, we can learn an enormous amount about the laws of physics that our universe obeyed in its first moments. Furthermore, the higher the velocity—or more precisely, the higher the energy—of the collisions we study with these machines, the closer they approach the conditions of the Big Bang itself. For example, in the era that preceded the formation of the first atomic nuclei—about a hundredth of a second or so after the Big Bang—the temperature of our universe was about 100 billion degrees. At this temperature, protons and neutrons typically zip around and smash into each other at speeds of tens of millions of meters per second—or roughly 10 percent of the speed of light. In the 1930s and 1940s, physicists began to use early particle accelerators—called cyclotrons—to study what happens when protons and other kinds of particles collide at such high speeds. Machines such as these have allowed us to learn how various forms of matter and energy behave under the high-temperature conditions that existed a hundredth of a second after the Big Bang.

Since the first cyclotrons, particle accelerators have become dramatically more powerful, allowing us to recreate and study the conditions that were present at even earlier times, and much closer to the Big Bang. The protons accelerated by the LHC carry with them roughly a million times as much energy as the particles studied at early cyclotrons, and thus they enable us to investigate much earlier eras of our universe's history. In the collisions observed at the LHC, a huge variety of particles can be created, including each of the seventeen forms of matter and energy described by the Standard Model. These kinds of interactions filled the early universe with a hot, dense soup of quantum particles. These particles interacted with each other constantly and were repeatedly created and destroyed. By studying the collisions observed at the LHC, we learn not only how matter and energy behave in our world today, but also how it behaved billions of years ago, only a trillionth of a second after the Big Bang.

One can't really talk about the first moments of our universe's history without talking about the exotic forms of matter that filled all of space during that era. To understand these first moments of time, we need to have a feel for the kinds of quantum particles that were present during the various epochs of cosmic history.

Let's start by imagining something simple—a box of electrons. The box is perfectly insulated, so none of the electrons can escape, and no energy or anything else can get in or out. If this were a box of everyday objects—like baseballs—you could feel confident that the number of objects in the box would remain fixed. They might bounce around, redistributing their energy and momentum, but a box of ten baseballs remains a box of ten baseballs.

Electrons, however, are not baseballs. Our intuition for objects like baseballs is the result of millions of years of natural selection, along with our own accumulated experiences. Even if you've never taken a physics class, you can look at a baseball as it is hit toward right field and make a pretty good estimate of where it will land and when it will get there. Our physical intuition is powerful. Nevertheless, there is little reason to think that it should prepare us to understand the behavior of electrons or other quantum particles. Our prehistoric ancestors didn't hunt with electrons, and we didn't grow up playing with quarks. Without this kind of trustworthy intuition to rely on, quantum particles can seem strange to us—they rarely behave the way we expect them to.

Our intuition does the best—or at least not the worst—when imagining a box of low-temperature electrons. In this case, the total number of electrons in the box will remain fixed. So at least in this respect our intuition gets things right. But even at low temperatures, electrons can radiate and create new particles of light—photons. Thus a box that starts out containing only electrons will, over time, evolve into a box of both electrons and photons.

The reason that the number of electrons stays fixed in this case is that each possesses electric charge, and, as far as we know, electric charge can never be created or destroyed. If the number of electrons in the box were to increase or decrease, the total amount of electric charge would change too. Photons, in contrast, carry no such electric charge, and thus can be freely created or destroyed.

In a higher temperature box—one containing particles with more energy—things can get a bit more interesting. According to Einstein's equation $E = mc^2$, two photons with enough energy can collide and convert their energy into mass, creating entirely new particles of matter. In such a collision, the incoming photons simply disappear and are replaced by a pair of new particles—for example, an electron and its antimatter counterpart, a positron. It's like a pair of colliding marbles that suddenly disappear, only to be replaced by a pair of baseballs, or a baseball and an anti-baseball if we take the analogy further.

In most respects, a positron is very much like an electron. It has the same mass as an electron and also carries electric charge. The key difference is that whereas an electron has a negative electric charge, the electric charge of a positron is positive. Every kind of particle in nature has an antimatter counterpart—a version of itself with the same amount of mass, but with opposite values for its electric charge and other quantum properties. The proton has the antiproton, the muon the antimuon, and so on. The only exceptions are those particles with no quantum properties to change the sign of. The photon, for example, has no electric charge or other such quantities. For this reason, the antimatter counterpart of the photon is simply the photon itself.

So in our high-temperature box, not only can the number of photons change, but also the numbers of electrons and positrons. The conservation of charge requires only that the number of electrons *minus* the number of positrons stays fixed. So an electron can be freely created or destroyed, but only if a positron is created at the same time. And electrons and positrons are not alone in this respect. If we turn up the temperature of our box even more, other kinds of particles will begin to be created—starting with pairs of muons and antimuons and working up to Higgs bosons and top quarks.

To understand how a collection of quantum particles will evolve with time, we need to understand how such particles interact with each other. When two baseballs collide, they bounce off of each other, but remain the same two baseballs. When two photons collide, two photons might come out, or three, or sixteen. And if the colliding photons have enough energy, they can transform into other kinds of particles—from electrons and positrons on up. Quantum particles do not behave like everyday objects. They follow very different rules than those favored by our intuition. Yet they do follow rules. By building particle accelerators and other experiments, we have discovered many of these rules, which are applicable across a wide range of temperatures and other conditions. From this information, we have been able to reconstruct how the particles in the box that we call our universe have behaved and transformed as space expanded and cooled over time.

A trillionth of a second after the Big Bang, the entire universe was filled with an unimaginably hot and dense plasma of energy. Throughout all of space, the temperature was greater than 10^{16} degrees—a billion times hotter than the core of the Sun. Conditions such as these are found almost nowhere in our universe today. But they are precisely the kinds of conditions that we are able to create and study at the LHC.

		Mass	Electric Charge	Color
Leptons	Electron	$0.0054\ m_{proton}$	Yes	No
	Muon	$0.11\ m_{proton}$		
	Tau	$1.9\ m_{proton}$		
	Electron Neutrino	$< 10^{-10}\ m_{proton}$	No	No
	Muon Neutrino	$< 10^{-10}\ m_{proton}$		
	Tau Neutrino	$< 10^{-10}\ m_{proton}$		
Quarks	Up	$0.002\ m_{proton}$	Yes	Yes
	Down	$0.005\ m_{proton}$		
	Strange	$0.10\ m_{proton}$		
	Charm	$1.4\ m_{proton}$		
	Bottom	$4.5\ m_{proton}$		
	Top	$185\ m_{proton}$		
Bosons	Photon	0	No	No
	Gluon	0	No	Yes
	W Boson	$86\ m_{proton}$	Yes	No
	Z Boson	$97\ m_{proton}$	No	No
	Higgs Boson	$133\ m_{proton}$	No	No

The seventeen fundamental forms of matter and energy that make up the Standard Model of Particle Physics.

So what does the LHC tell us that our universe was like in this primordial state, only a trillionth of a second after the Big Bang? First of all, all known forms of energy and matter—every particle in the Standard Model, including all of the forms of antimatter—were present and in high abundance. This includes not only photons and electrons, but also many other less-familiar particle species. For example, there exist particles—muons and taus—that are a lot like electrons, but that are heavier and very unstable. In addition, there are three ultra-light and electrically neutral particles called neutrinos. These neutrinos, along with the electron, muon, and tau, are collectively known as the leptons. The six leptons are mirrored by a group of six other particles, called quarks. In addition to electric charge, the quarks carry another kind of charge, called "color." This kind of color has nothing to do with the visible qualities of light, however. Whereas the electric charge of a particle causes it to feel the effects of the electromagnetic force, a particle's color enables it to experience the strong nuclear force—the force that binds quarks together to form protons and neutrons, and that holds protons and neutrons together inside of atomic nuclei.

The ultra-hot primordial plasma that once filled all of space contained all of the leptons, quarks, and other known particle species and in utterly astonishing quantities. The density of energy in this plasma was equivalent to more than 10^{36} kilograms in every cubic meter of space. To obtain a density this high, you would have to compress the entire Sun into a region the size of a marble, or the Earth into a sphere a tenth of a millimeter across. Under these ultra-hot and high-density conditions, every particle was constantly being bombarded and smashing into others. In some of these interactions, two particles would simply scatter and recoil off of each other. But in many others, the incoming particles would be destroyed and converted into other forms of energy. These kinds of interactions happened so frequently during this era that no individual particle survived for very long. Within even a fraction of a trillionth of a second, the energy possessed by a given particle would change forms many trillions of times. Energy in the form of an electron might be converted into a photon, and then into a Higgs boson, followed by a top quark, transforming over and over again. Nothing was permanent in this era. Everything was fleeting and in flux.

During these first moments of our universe's history, space was expanding at a staggering rate. Between a trillionth of a second after the Big Bang and a billionth of a second later, the volume of our universe increased by a factor of about 30,000, and the temperature dropped by a factor of 30. In any given moment—any fraction of a blink of an eye—everything was almost instantly transformed. Space itself was in a state of continuous and relentless change.

As space expanded and cooled, some types of particles began to disappear from our universe. The first of the known particles to become scarce was the top quark—a particle that was discovered in 1995 at the Fermi National Accelerator Laboratory, where I conduct most of my own research. Of all of the known fundamental particles, the top quark is the heaviest. A single top quark has about 185 times as much mass as a proton, or about 360,000 times that of an electron. In the collisions that were taking place a trillionth of a second after the Big Bang, there was usually enough energy for the incoming particles to transform into a top quark and an anti–top quark. But by a hundred trillionths of a second or so, the temperature had dropped low enough so that top quarks began to disappear more often than new ones were created. By a billionth of a second after the Big Bang, top quarks, Higgs bosons, Z bosons, and W bosons had all become scarce throughout our universe.

As time went on, the composition of the particles found throughout our universe continued to change and evolve. This began with the disappearance of the heaviest forms of matter, but other changes would soon follow as well. For example, up to this point in time, quarks and gluons had been free particles. By this I mean that a given quark or gluon would move through space on its own, interacting with other forms of matter and energy just as any other particle might. But around ten millionths of a second or so after the Big Bang, these particles began to find themselves becoming irresistibly attracted to one another. The era of the free quark was coming to an end. Within a fraction of a millisecond, all of the quarks and gluons had become bound together into small groups, forming composite objects such as protons, neutrons, and pions.

This transition from a plasma of free quarks and gluons into a gas of bound particles was dramatic, and it left our universe in an almost unrecognizable state. Like other eras of the early universe, much of what we know about this transition we learned from particle accelerators. To recreate the conditions of the early universe's quark-gluon plasma, physicists do not generally use protons, but instead accelerate and collide large nuclei, such as lead or gold. By accelerating such large nuclei, physicists can effectively collide hundreds of protons and neutrons together simultaneously, creating miniature fireballs filled with an incredibly high density of particles and energy. Physicists at the LHC, as well as at the Relativistic Heavy Ion Collider (RHIC) in New York, have indeed been able to create tiny fireballs filled with quark-gluon plasma. The temperature within these fireballs is about four trillion degrees—similar to that of our entire universe only two hundredths of a millisecond after the Big Bang.

Accelerator experiments have taught us a great many things about how matter and energy behaved in the early universe. From the results of these experiments, we have mapped out many of the epochs of our universe's history, going back as far as a trillionth of a second after the Big Bang. But despite all of their utility, these experiments also have their limitations. Significant events may very well have taken place in the early universe that we don't know about yet—events that are hiding in the blind spots of the LHC. There may be forms of matter and energy that we have not yet discovered but that were commonplace in the first moments after the Big Bang. Accelerators can and have taught us many things about our universe's early history. But they do not necessarily reveal the entire story.

The LHC collides particles together with enormous amounts of energy. In this respect, it really does recreate the conditions of the early universe. But there are other ways in which the environments created at the LHC do not even come close to mimicking those of our earliest eras. Perhaps the most important of these distinctions has to do with the fact that we are able to study only a relatively modest number of collisions at the LHC.

Now I appreciate that the word "modest" might seem like a strange choice in this context. After all, I am talking about a machine that collides 700 million protons together every second, adding up to around 10 quadrillion—or about 10^{16}—pairs of protons to date. By most reasonable measures, this is not a modest sum. But it is absolutely tiny when compared to the number of collisions that occurred among particles in the early universe.

Imagine, for example, an individual electron that was present in our universe a trillionth of a second after the Big Bang. Under the conditions of that era, it will be only about 10^{-30} seconds or so, on average, before this electron interacts with another particle, potentially causing it to transform into another kind of particle. The new particle will then go on to collide and possibly transform again, only 10^{-30} seconds or so after that. At this rate, about 10^{18} interactions take place per particle, all within a window of only a trillionth of a second. This is more than all of the collisions that have ever taken place at the LHC put together. This means that there could plausibly exist kinds of interactions that were relatively common in the early universe, but that are at the same time so rare that we will likely never witness them at the LHC. Such interactions could have played an important role in the shaping of our universe, impacting not only the forms of matter and energy that were once present, but also potentially influencing how space itself expanded and evolved.

Every form of matter and energy—from electrons to Higgs bosons—has had an impact on our universe's history. If any one of the known particles did not exist, our universe would have expanded and evolved differently, in particular during the first moments that followed the Big Bang. On the other hand, if there exists a particle that we have *not* yet discovered, it would almost certainly have had an impact on our universe's first moments. For this reason, cosmologists often find themselves imagining—proposing or hypothesizing—new kinds of particles and exploring their consequences for our universe.

In this spirit, let's imagine for a moment that there exists an unknown kind of particle—one that we have never detected at the LHC or at any other experiment. And let's even give this new particle a name—the *mysterion*. Accelerator experiments tell us that when particles collide, they can be converted into other forms of matter—quarks, photons, neutrinos, and so on. Depending on how mysterions interact with other kinds of particles, it's possible that they too could be created in a small fraction of such collisions. But in order for mysterions to have gone undetected at the LHC, the interactions that lead to their creation must be very rare. It might be hundreds of years before the first mysterions are created at the LHC, for example. But even if this were the case, the same kinds of interactions would have occurred a multitude of times in the early universe. And with every such interaction, a small quantity of the universe's energy would have been transferred into a growing sea of mysterions.

If these new particles were anything like those we already know, they would interact repeatedly with the surrounding particles in the early universe, quickly reaching a state of equilibrium and being created at the same rate as they are destroyed. But since the interactions that lead to the creation of mysterions are so rare, its other interactions with the known forms of matter will be rare as well. So instead of acting as part of the plasma of constantly and continuously interacting particles, the mysterions might very well behave more like ghosts—passing through space without interacting much at all with the surrounding sea of matter and energy.

The Standard Model contains three examples of particles that act a lot like ghosts—the neutrinos. Roughly 100 trillion neutrinos pass through your body every second of every day, most of which were produced in the Sun as a by-product of nuclear fusion. These ultra-light particles have virtually no impact on your body or your health, because they interact so rarely. In fact, these particles can pass through the entire Earth without any appreciable effect. To neutrinos, it is almost as if matter doesn't exist.

This wasn't always the case, however. When the universe was very young and very hot, neutrinos were abundant throughout all of space, and they interacted frequently with many of the other known types of particles. It wasn't that neutrinos themselves were so different then than they are now, but rather that the density of particles that they could potentially interact with was so mind-bogglingly high. Just as they can now, most neutrinos could pass through the equivalent of many Earths' worth of matter before interacting. But a billionth of a second after the Big Bang, the density was so high that every cubic centimeter contained roughly the equivalent of an Earth's mass of energy. At such high densities, even the neutrinos were constantly bombarded by the sea of surrounding particles.

As the universe expanded and its density decreased, however, the neutrinos gradually interacted more and more rarely with the surrounding matter and energy. Eventually, about one second after the Big Bang, their relationship with the surrounding matter and energy ended. From that point in time forward, neutrinos have—on average—been able to move through space unimpeded, indifferent and unresponsive to the surrounding matter and energy. As space expanded over the past 13.8 billion years, these neutrinos steadily cooled, but little else occurred to these ghostly particles as they traveled through space. And just as our universe is filled with a background of photons—the cosmic microwave background—all of space contains a sea of low-energy neutrinos. But whereas the cosmic microwave background was formed 380,000 years after the Big Bang, these relic neutrinos are much older, having been traveling unimpeded through space since our universe was only one second old.

Let's return for a moment to our newly hypothesized mysterions. If these particles are stable—as neutrinos are—they may have survived until the present day, surrounding all of us and filling all of space. If they are heavy, these particles might make up some or all of our universe's dark matter. If they are lighter, on the other hand, they could form another of our universe's cosmic backgrounds, alongside those consisting of photons and neutrinos. In modern times, these kinds of cosmic backgrounds have had little impact on our universe's evolution, but this was not always the case. According to the equations derived from general relativity, the more energy that is present in space, the faster that space expands. If a population of light mysterions survived the Big Bang, the energy stored in these particles would have caused our universe to expand faster than we otherwise would have expected, especially during the first hundred thousand years or so of cosmic history.

To test whether mysterions or other such particles might have been present during our universe's first hundred thousand

years, cosmologists measure and study the detailed characteristics of the cosmic microwave background. The information
contained in the patterns of this light can be used to determine
how fast the universe was expanding during its first few hundred thousand years, and thus indirectly measure how much
energy was present in the form of photons, neutrinos, and other
light or fast-moving particles. Our most recent and precise measurements indicate that the total energy in these kinds of particles was quite close to what we expected to be present in the
form of photons and neutrinos—within 5 percent or so of
the predicted value. In other words, although there may have
been some mysterions present during this era, they didn't make
up very much of the total energy.

As I am writing this, observational cosmologists are working
hard to design and build new experiments that will be able to
measure the cosmic microwave background in even greater detail. The current projection is that these efforts will improve the
precision of this measurement by a factor of about 10. This
means that if there are no unknown light or fast-moving particles that contribute to the energy density of the universe, these
experiments should end up measuring a value that is within
about 0.5 percent of that predicted from photons and neutrinos
alone. But if any particles like our mysterions do exist—hiding
in the blind spots of the LHC—these upcoming observations
have a very good chance of detecting their subtle influence on
our universe's evolution and history.

The LHC and other particle accelerators have taught us a
great deal about our universe's early history. But they cannot tell
us everything that we might want to know about the forms of
matter and energy that filled all of space during the first fraction
of a second after the Big Bang. Perhaps most of all, the LHC
tells us little to nothing about our universe's first trillionth of a

second. Only with a larger and more powerful accelerator will we be able to explore these earliest moments of time. And although it has been profoundly informative in any number of ways, the LHC has not yet enabled us to discover the nature of dark matter or to resolve the mystery of why our universe contains so much matter and so little antimatter. The puzzles of dark energy and inflation remain just as baffling as they were on the day the LHC was turned on. In the face of these cosmic mysteries, we continue to build and operate new accelerators and new telescopes to push farther back in time, exploring earlier eras and looking deeper into the first moments of cosmic history.

6

The Origins of Everything

I . . . a universe of atoms, an atom in the universe.

—RICHARD FEYNMAN

OVER THE PAST few decades, cosmologists have discovered that most of the energy in our universe takes the form of dark matter and dark energy. Yet these substances have little to do with the world of our ordinary experience. We cannot touch dark matter, and we cannot see dark energy—at least not in any direct way. And we ourselves are not made of these mysterious materials. Instead, everything we directly experience consists of one thing and one thing alone: atoms.

Although much of what we know about atoms falls within the realm of chemistry, cosmologists too spend their time pondering these basic building blocks of matter. But whereas chemists study how these objects behave and bind together to form molecules, cosmologists are more concerned with how they came to exist in the first place. As it turns out, there are few questions about our universe that are more perplexing than how and why the fundamental constituents of atoms—quarks and electrons—came to exist.

Every cosmology textbook has a chapter or more on the creation of quarks and electrons in the first moments after the Big Bang. And at first glance, the calculations involved appear to be fairly straightforward. We know how to calculate how many quarks were present in the early universe and how this quantity changed as our universe expanded and cooled. We can also calculate when these quarks coalesced in groups of three, to form the first protons and neutrons. This should not be a hard problem.

Yet it is. When we carry out these calculations, we find that they simply do not predict that a universe like ours—one full of atoms—should have emerged from the Big Bang. Instead, they tell us that very few quarks and very few electrons should have survived these first moments. According to the math, the world should be virtually free of atoms.

The resolution to this problem is not that we have made a simple error in our calculations—no one has forgotten to carry the 1. Instead, the fact that our universe contains large numbers of atoms is proof that our understanding of the Big Bang is incomplete. During its first fraction of a second, our universe must have contained forms of matter and energy that we have never observed in any experiment. And it must have been host to significant transformative events—events that we still know almost nothing about. We know this is true because if it weren't, we would not exist.

Using particle accelerators such as the LHC, physicists have learned what our universe was like when it was very young and very hot—filled with a soup containing all of the known particle species. This included not only the particles of matter—electrons, quarks, neutrinos, and so forth—but antimatter as well— positrons, antiquarks, antineutrinos. From what we currently understand, matter cannot be created without creating an equal amount of antimatter. And whenever matter is destroyed, antimatter is destroyed along with it. The fates of these two substances are closely intertwined.

This relationship between matter and antimatter makes it hard for us to understand the world that we see all around us. After all, if our universe was once filled with both matter and antimatter, and these substances are always created and destroyed in conjunction with each other, then we have to ask ourselves: How did we come to be living in a world that is filled with so much matter, while containing so little antimatter? Where did all of the antimatter go—and why didn't all the matter go with it?

From what we have learned at the LHC and other experiments, we would expect that pairs of quarks and antiquarks— along with other matter-antimatter pairs—should have disappeared almost entirely from our universe over the course of our universe's first fraction of a second. According to what we know about our universe and its physical laws, matter-antimatter pairs should have been destroyed far more often than new pairs were created, thereby ensuring that essentially none of these particles—matter or antimatter—would survive the heat of the Big Bang. Yet some did. Somehow, some mechanism or event—something unknown to us—must have stepped in to prevent this outcome. More specifically, something must have happened in that primordial soup that caused the total amount of matter present to become slightly greater than the total amount of antimatter. Even a very small imbalance, if established early enough in our universe's history, could explain the world we see all around us today. A primordial asymmetry as slight as 10 billion and 1 particles of matter for every 10 billion particles of antimatter would be enough to enable a small but adequate fraction of the matter to survive and escape the early universe intact.

Although we're still trying to figure out how this small imbalance between matter and antimatter came to exist, the fact that it somehow did gives us an important hint about the particles that must have been present and the events that must have taken place in the first fraction of a second after the Big Bang. If it weren't for this unknown and mysterious mechanism, our universe would contain essentially no atoms. And without atoms, there would never have been any gas or dust—or galaxies, stars, or planets. Our universe would contain no chemistry. No life. And no us.

The relationship between matter and antimatter is deeply embedded into the structure of our universe—the quantum nature of our world makes sense only with both matter and antimatter included within it. This symmetry is at the heart of our universe and its physical laws.

Over the past half-century or so, theoretical physicists have discovered just how powerful the concept of symmetry can be, using it to construct theories that have successfully predicted many of the kinds of particles that would later be created and discovered at accelerator experiments. With near consensus, they described the expected properties of the Higgs boson, the top quark, the W and Z bosons, and several other particles long before they were observed in any experiment. To these physicists, the mathematical structure behind their theories simply requires the existence of these new forms of matter. And although the theories that make up the foundation of modern particle physics cannot address every question that we would like to ask about our universe, they have repeatedly enabled us to predict previously unobserved phenomena whose existence would only subsequently be confirmed by experiments. These theories have proven to be very powerful indeed.

This is very different from how things were throughout most of the history of science. Newton didn't predict that gravity *must* exist from some overarching mathematical argument. Instead, he showed that an attractive force of gravity could explain already existing observations of both planetary motions and falling bodies. Similarly, Einstein never claimed that light waves logically *had* to take the form of individual photons; he merely argued that the observed photoelectric effect could be explained if this were the case. These physicists and many others strove to find new and more powerful ways of explaining the results of previously conducted experiments. They only rarely predicted entirely new phenomena before they were observed.

In my reading of the history of science, it was around 1928 or so that physics began to depart from this trend. It was in that year that a young theoretical physicist named Paul Dirac recognized for the first time something that was hidden within the very mathematics of the then-new theory of quantum physics. To Dirac, the equations made it clear that in order for electrons to exist, there must also exist another kind of particle—a particle a lot like an electron, but with a positive electric charge. Although such particles had never been observed or detected in any experiment, their existence appeared to be an unavoidable consequence of the very structure of our universe. Before Dirac, no one had ever successfully predicted the existence of a fundamental constituent of matter before it had been observed; the effects of the electron, proton, neutron, and photon were each measured in experiments well before anyone argued that they must exist. Breaking from this trend, Dirac boldly predicted the existence of an entirely new form of matter, without having seen it or even its effects. He predicted the existence of what we now call antimatter.

In 1932, only four years later, positrons—the antimatter counterpart of the electron—were observed and measured for the first time, confirming that Dirac was indeed correct. And in the 1950s, physicists used particle accelerators to create and observe both antiprotons and antineutrons for the first time.[1] Today, we recognize that all forms of matter have antimatter counterparts. This relationship is a fundamental symmetry built into the foundation of our quantum world. The very logical structure of our universe requires the existence of both matter and antimatter.

Antimatter has long fascinated and confounded cosmologists. Everything we know about this mysterious stuff places it on an equal footing with matter. Antiparticles possess the same mass, are created and destroyed in the same ways, undergo the same interactions, and otherwise represent seemingly perfect—and yet opposite—copies of their particle counterparts. It is the fact that these substances are so precisely alike that makes it so hard for us to understand why there is so much more of one than the other in our universe today.

Thus far, I've been asserting that there is very little antimatter in our universe, as if this were an entirely obvious fact. But how

---

1. You might wonder how the neutron—being electrically neutral—could have an antimatter counterpart. After all, won't the neutron and antineutron both have zero electric charge? The difference is that whereas a neutron is made up of three quarks, with charges of $-1/3$, $-1/3$, and $+2/3$, an antineutron is made up of three antiquarks, with charges of $+1/3$, $+1/3$, and $-2/3$.

sure of this are we really? Here on Earth, we can be quite certain that antimatter is extremely rare—and thankfully so. Any time that antimatter and matter are brought into contact with each other, they instantly obliterate each other, releasing raw energy in their wake. Although individual particles of antimatter—like those created at the LHC—are harmless enough, any macroscopic quantities of this substance would be ravenously destructive and simply couldn't go unnoticed. The presence of a single gram of antimatter, for example, would generate an explosion in excess of 20 kilotons of TNT—comparable to the energy released in the nuclear detonations at Hiroshima or Nagasaki.

But even if there is almost no antimatter near Earth, might there be large quantities of it elsewhere in our universe? This possibility is much harder to dismiss. In the decades that followed the discovery of antimatter, many physicists considered it likely that our universe contains equal amounts of both matter and antimatter. After all, just because we happen to live in a region that is overwhelmingly dominated by matter doesn't preclude the existence of other regions of space that are instead dominated by antimatter. And if such regions do exist, then perhaps our universe as a whole might contain the same quantities of matter and antimatter, neatly resolving the mystery at hand.

If these antimatter regions exist, though, what might they look like, or be like? In isolation, antimatter behaves just like ordinary matter does. In some region of our universe—very far away from us—there could be large quantities of antiprotons, antineutrons, and positrons without the presence of any protons, neutrons, or electrons to annihilate with. These antiparticles would combine to form antimatter versions of all of the known types of atoms and molecules, which would undergo the full array of physical and chemical reactions and processes that take place among matter. The antimatter in such a region could form stars, planets, and galaxies—and could even contain life. And the light generated from an antimatter star or an antimatter galaxy would be entirely indistinguishable from that produced by any ordinary star or galaxy.[2] The next time you look upon the night sky, you might wonder whether some of that light might have been generated in the nuclear furnace of an antimatter star. In perfect isolation, there would be no way to know for sure.

In practice, however, there is no perfect isolation. No real star or galaxy exists entirely apart from others. Astrophysical systems pass matter back and forth constantly. At any given moment, there are many galaxies in our universe that are colliding or merging with one another. Even our own galaxy is expected to collide with its nearest neighbor, the Andromeda galaxy, in another 4 billion years or so. If a galaxy made up of matter were to collide with an antimatter galaxy, it would generate the most dramatic and destructive event since the Big Bang. Needless to say, such events—if they ever take place—must be incredibly rare.

On the far less extreme side, astrophysicists have used X-ray and gamma-ray telescopes to search for the signatures of more modest quantities of antimatter that might be interacting and annihilating with matter. But to date, they have found no evidence of any significant reservoirs of antimatter in our universe. Our universe is truly dominated by matter, accompanied by only the smallest traces of antimatter. If antimatter stars or

2. Unlike most other kinds of particles, the photon has no electric charge or other such properties, making it its own antiparticle. The opposite of zero is zero, and thus an antiphoton is exactly the same thing as a photon.

antimatter galaxies exist at all, they must be extremely rare and very far away.

The fact that there is far more matter than antimatter in our universe flies in the face of everything physicists have learned from accelerator experiments about how both matter and antimatter behave. For reasons that we do not yet understand, these substances managed to avoid their mutual annihilation in the early universe. Somehow, much more matter than antimatter survived the first moments that followed the Big Bang.

But how did this happen? What took place in the first fraction of a second of our universe's history that made the existence of matter—along with stars, planets, and life—possible? Frankly, we just don't know. This is an open question, and one of the most fascinating and important in all of cosmology.

To be clear, it's not that we don't have any ideas for how our universe might have come to be dominated by matter over antimatter. We have lots of these ideas. Like many other theoretical cosmologists, I've occasionally written a paper that proposes how this might have taken place. The scientific journals are filled with thousands of such papers. The problem is that we have no way to tell which, if any, of these ideas is right. We're like a group of detectives making guesses about the criminal behind an unsolved case—maybe it's Joe Smith, or Katie Jones, or countless other suspects. It's easy to propose a suspect. But without evidence, it's impossible to make a solid case.

When it comes to how matter emerged over antimatter in the early universe, our list of suspects is long, and our evidence for any one of these hypotheses over another is scant. Yet all of these ideas have at least one thing in common. They were all built upon the work of a Soviet physicist, activist, and nuclear architect named Andrei Sakharov.

Like many great scientists, Sakharov's career did not follow the straightest of paths. After completing his PhD in theoretical physics in 1947, he began work on the Soviet Union's nuclear weapons program, making significant contributions to the first Soviet nuclear weapon—a plutonium device similar to the bomb dropped on Nagasaki. But this was not enough to achieve parity with the Americans, and Sakharov rose to become the single most important figure in the Soviet effort to develop even more destructive weapons, culminating in the detonation of the 50-megaton "Tsar Bomba" in 1961.

Like many other scientists working on nuclear weapons during this era, Sakharov became increasingly concerned about the ethical and political ramifications of his work. In the late 1950s he became politically active, advocating against nuclear proliferation and for the end of atmospheric nuclear tests, and playing a significant role in the signing of the 1963 Partial Test Ban Treaty. To Sakharov, the existence of nuclear and thermonuclear weapons required great care and caution—more than either superpower was exhibiting. History is littered with examples of human foolishness. But this was different. For the first time, humankind was becoming capable of making its last mistake.

For these reasons and perhaps others, Sakharov decided to redirect his scientific efforts toward more peaceful purposes, setting out to reinvent himself as a particle physicist and a cosmologist. By this time, however, Sakharov was in his mid-forties and had been out of touch with the latest developments in theoretical physics for nearly two decades. Most of his colleagues didn't even think of him as a theoretical physicist at all, but as an inventor or perhaps an engineer. The odds that someone in this situation would ever be able to make significant contributions to fundamental physics seemed slim, at best. No one expected that Sakharov was about to revolutionize the way we think about our universe's first moments.

As Sakharov was making his return to physics in the mid-1960s, cosmology was coming of age. The discovery of the cosmic microwave background in 1964 had elevated the Big Bang Theory from what had once been perceived as a rather dubious and speculative idea to a respected pillar of the scientific mainstream. And during this same period of time, the field of particle physics was thriving as well, basking in the newly appreciated power of what are known as gauge theories. Within a period of only a few years, this class of theories was used to predict the existence of both quarks and the Higgs boson, and provided the basis for the construction of the Standard Model of Particle Physics. For both cosmology and particle physics, it was a golden age.

As these two scientific communities were making their respective advances, however, they continued to operate almost entirely independently of one another. Most cosmologists knew little about the developments taking place in particle physics, and a typical particle physicist was more or less unaware—and perhaps skeptical—of the progress that had been made in cosmology. If you were a particle physicist, you probably didn't read any papers or attend any seminars about cosmology, and vice versa. If I had to speculate, I'd guess that this rift played a big part in what made Sakharov so successful during this era. Having been sequestered from the physics community for two decades, he didn't have the same preconceptions about what he should or shouldn't be interested in or be working on. As a result, Sakharov was able to produce and publish papers on the new theory of the strong nuclear force—known as quantum chromodynamics—as well as on general relativity, and on how distributions of matter evolve as space expands. He wasn't a particle physicist, and he wasn't a cosmologist. It was precisely this that made it possible for him to become both.

Sakharov's single most important contribution to fundamental physics was described in a paper that few particle physicists or cosmologists at the time could have imagined writing. As a particle physicist, he appreciated the symmetry between matter and antimatter that was built into the fabric of our universe. And like other cosmologists, he understood that under the ultra-dense conditions of the Big Bang, matter and antimatter would eradicate each other, leaving nothing behind to form the atoms found in our universe. Although equal amounts of matter and antimatter should have initially been present after the Big Bang, it was clear that somehow our universe managed to transition from that state into one that contained more matter than antimatter. More than anyone else, it was Sakharov who laid the groundwork to explain how a transition of this nature may have taken place.

In a short paper published in 1967, Sakharov laid out three conditions that he demonstrated must be met in order for a perfectly balanced collection of matter and antimatter to become dominated by matter. Although they can be complicated to describe in their full technical glory, these conditions can be paraphrased roughly as follows:

1. There must exist some kind of interaction between particles that can change the total number of quarks minus the total number of antiquarks.
2. Nature must have a bias in these interactions that favors the creation of matter over antimatter, or the destruction of antimatter over matter.

3. These interactions must have had an opportunity to act under particularly violent or rapidly changing conditions at some point in the early universe.

What should we make of Sakharov's conditions, and what do they tell us about our universe? The fact is, there are any number of ways that they might have been satisfied during our universe's first fraction of a second, and we have little information to distinguish between these many possibilities. That said, Sakharov's conditions still provide us with a valuable insight into the kinds of interactions that must have taken place and the conditions that must have been present during our universe's earliest moments.

In the years that followed the publication of Sakharov's greatest scientific paper, few other physicists took much notice. In its first ten years, it was cited only a handful of times. But beginning in the late 1970s, more and more particle physicists and cosmologists alike began to appreciate the profundity of Sakharov's question, as well as the hints that his three conditions give us about its answer. In the decades that followed, thousands of papers expanded upon these insights, often proposing specific ways in which Sakharov's three conditions might have been met. Physicists now use the word baryogenesis—literally the origin of baryons, which are made of quarks—to refer to the transition from a balanced state of matter and antimatter to one dominated by matter. Today, we still don't know exactly how baryogenesis took place in our universe. But we do know that the three conditions laid out by Sakharov half a century ago were somehow satisfied. This fact provides us with some of our most valuable clues about how our universe evolved in its first moments and the forms of matter and energy that were present during that primordial era.

* * *

Of Sakharov's three conditions, the first is the most straight-forward to understand—some might even say obvious. To transform a world that contains equal numbers of quarks and antiquarks into one with only quarks, interactions that create a net number of quarks—or destroy a net number of antiquarks—must be at play. That said, no interactions that accomplish this have ever been observed, and in Sakharov's time, many physicists took it for granted that no such interactions were possible. For one thing, if quarks can be destroyed without destroying antiquarks, then it should also be possible for particles such as protons—which are made of quarks—to decay. In other words, Sakharov's first condition implies that every atom in our universe is ultimately unstable—at least slightly. Even atoms cannot last forever.

Interestingly, another argument was proposed around the same time that also seemed to imply that quarks and antiquarks must be able to be created and destroyed asymmetrically. This argument has to do with black holes, and what has become known as the "no-hair theorem." In a nutshell, this theorem says that you can completely describe a given black hole by only three quantities: its mass, its electric charge, and its rate of rotation. This means that any two black holes with the same mass, charge, and spin are completely identical in every respect. From this it follows that an electrically neutral black hole that forms out of matter is exactly the same as one that forms out of antimatter. The black hole's gravity essentially erases the net number of quarks that went into its formation. And although physicists still debate the full consequences of this theorem, there is broad agreement that it provides support for the idea that nature satisfies the first of Sakharov's three conditions.

To this day, no process that changes the net number of quarks has ever been observed. Powerful experiments have been designed to search for instances of proton decay, but they have all come up empty handed. Instead, these experiments have informed us that the proton's half-life is at least 10^{34} years, implying that no more than one proton out of 10^{24} has decayed over our universe's 13.8-billion-year history. But despite the lack of evidence, most contemporary particle physicists are reasonably confident that protons do not live forever—even if they do persist for a very, very long time. Efforts to more stringently test the stability of the proton continue, such as those planned for the DUNE experiment in South Dakota's Homestake Mine. Perhaps this experiment will finally observe the decay of the proton, enabling us for the first time to place an expiration date on the atoms and molecules found throughout our universe.

At first glance, Sakharov's second condition might also seem fairly straightforward. After all, if you can imagine an interaction that reduces the number of antiquarks relative to quarks, you can easily imagine another that does the opposite. In order for the effects of these different interactions to avoid canceling each other out, there must be a preference for one over the other. Nature must somehow be biased against antimatter.

The presence of such a bias within the laws of physics would carry deep implications for the natures of both space and time. In particular, it would reveal things about the symmetries that are built into the very foundations of our universe. When physicists say that something exhibits a symmetry, they mean that there is something you can do to it that will leave it unchanged. Take a perfectly uniform sphere, for example. Such a sphere possesses a rotational symmetry, because you can change its orientation by any angle without altering it in any way. An irregularly shaped rock, in contrast, has no such symmetry. Symmetries are at the heart of modern physics. And few symmetries are as fundamental or as far-reaching as those associated with the concepts of parity, charge, and time.

Parity symmetry is a symmetry of space. Anything that is indistinguishable from its mirror image is symmetric with respect to parity. A perfect cylinder, for example, looks exactly the same as its reflection, and thus is parity symmetric. Carve the cylinder into the shape of a screw, however, and you break this symmetry—the mirror image of a clockwise winding screw is a counterclockwise winding screw.

To understand what I mean by charge symmetry, imagine a collection of protons and electrons. If I could somehow instantly replace all of the protons with negatively charged antiprotons and at the same time replace all of the electrons with positively charged positrons, each particle would feel exactly the same electromagnetic force as it did before the swap. The antiprotons and positrons still have opposite charges from each other, and so attract each other in exactly the same way as they did before the change took place. This is because the electromagnetic force is perfectly symmetric with respect to charge.

Lastly, something is symmetric with respect to time if it doesn't distinguish a process moving forward in time from one moving backward in time. If I show you a video of a baseball flying upward, reaching the top of its arc and falling back towards Earth, and then play you the same video backwards, you won't be able to tell which version of the video is which. The force of gravity doesn't distinguish between the two—it is time symmetric.

If I didn't know better, I would have guessed that our universe's physical laws would essentially have to be symmetric with respect to parity, charge, and time. It just seems intuitive that nature *should* be symmetric in these ways. With a perfectly symmetrical top, for example, I feel like it should work the same way regardless of whether I spin it clockwise or counterclockwise. But nature, it seems, does not care what my intuition says. All three of these symmetries are indeed gently broken by the physical laws of our universe.

Of the four known forces, three of them—the gravitational, electromagnetic, and strong forces—act the same way on a given particle, regardless of whether it is spinning clockwise or counterclockwise. In other words, these forces are symmetric with respect to parity. But this is not the case for the weak force. In experiments beginning in 1956, it was shown that particles of matter feel the effects of the weak force only if they are spinning counterclockwise. A clockwise spinning particle of matter simply does not experience this force at all. Weirder still, this relationship is reversed for antimatter particles: they feel the weak force only if they are spinning clockwise. Through the eyes of the weak force, a particle is like a vampire—if you can see it, its mirror image is invisible.

The discovery that the weak force does not respect parity or charge symmetry overturned much about what particle physicists had assumed about their universe. They learned the hard way that the weak force does not conform to their naïve intuition. Like so many other times in the history of science, their common sense had led them astray. But in the wake of this discovery, they realized that even if the weak force breaks the symmetries of parity and charge, the *combination* of these two symmetries might remain intact. If I start with a particle that feels the weak force—a counterclockwise spinning electron, for example—and then simultaneously change both its orientation *and* its charge—making it a clockwise spinning positron—it continues to feel the weak force. So even though the weak force breaks the individual symmetries of charge and of parity, it seemed that at least this combined symmetry of both charge and parity—CP symmetry—might still be built into the structure of our universe.

But even that turned out not to be the case. In 1964, particle physicists were shocked once again when experiments showed that CP symmetry is broken as well. This phenomenon was observed for the first time among particles called kaons—states consisting of a strange quark and a down antiquark bound together. It was previously known that the weak force could transform a kaon into an antikaon, and vice versa. The new fact that these experiments revealed was that these processes do not happen with the same likelihood in both directions. There is, in fact, a bias between matter and antimatter built into nature's very structure. The violation of CP symmetry is exactly what is needed to satisfy Sakharov's second condition, and this makes it is possible—at least in principle—for processes that favor matter to win out over those favoring antimatter.

Perhaps strangest of all is what this realization tells us about the nature of time. Absolutely any quantum field theory that is compatible with Einstein's theory of special relativity must be symmetric with respect to the combination of charge, parity, and time—CPT symmetry. This means that if CP symmetry is broken by the weak force, then the symmetry of time must be broken as well. Unlike the arc of the baseball—which looks the same whether moving forward or backward through time—the weak force acts differently in these two temporal directions. In this way, nature distinguishes between the past and the future, making it possible to ever so gently tilt the scales in favor of matter over antimatter as time advances.

* * *

Even if particles can interact in a way that changes the net amount of matter relative to antimatter, and even if those interactions are biased in favor of matter, this alone cannot explain how so many protons, neutrons, and electrons managed to survive the heat of the Big Bang. If our universe steadily and continuously expanded and cooled over the course of its early history, then even these kinds of interactions would move the relative quantities of matter and antimatter only toward a natural state of equilibrium, ultimately failing to break the universe out of its primordial state of balance. In order for matter to emerge victorious over antimatter, Sakharov showed that a third condition must also be satisfied: our universe must have experienced a particularly violent transformation during its first moments.

There are a number of ways in which Sakharov's third condition might have been met. Our universe may have undergone a brief period of sudden expansion, during which the interactions favoring matter were allowed to take effect. Alternatively, there may have been an exotic species of particle—unknown to us today—that didn't interact with the other particles in the early universe, but created an abundance of new quarks as they decayed. Or, perhaps most interesting of all, our universe may have undergone a sudden, dramatic change at some point during its first second—not merely a steady cooling, but an abrupt transition, something analogous to water boiling and transforming into steam.

Examples of phase transitions are found throughout chemistry and physics. And although phase transitions between solids, liquids, and gases are the best known, there are others as well. In fact, we know that there were at least three phase transitions that took place in the early history of our universe. The most recent of these was the transition from free electrons and protons to bound atoms—a transformation that took place about 380,000 years after the Big Bang. Much earlier—only a fraction of a millisecond after the Big Bang—an analogous transformation occurred, when all of the free quarks throughout our universe became bound together to form protons and neutrons. And even earlier—around a trillionth of a second after the Big Bang—another kind of phase transition took place. Before this transition, Higgs bosons behaved much like any other particle. But as the temperature of our universe dropped below 10^{16} degrees or so, this changed, and a field associated with these Higgs bosons began to act on the other particles in a new and different way. For the first time, particles such as the W and Z bosons—and likely the quarks and leptons as well— began to slow down to speeds well below the speed of light. Before this transition, these particles had moved like light and had no mass. But the change in the Higgs field caused them to carry more inertia and to resist acceleration. In effect, it was this change in the Higgs field that gave these particles their mass. Without the presence of the Higgs field, all of the known fundamental particles would be massless, making our world entirely unrecognizable.

From what I've written in this chapter so far, it might seem that we have a satisfactory answer to Sakharov's conundrum. There is every indication that his first condition—the existence of interactions that change the net number of quarks relative to antiquarks—is satisfied in our universe. And his second condition—a bias in nature in favor of matter over antimatter—has been confirmed by numerous experiments carried out over the past half-century. Finally, we know of multiple phase transitions that took place in the early universe. It would seem that we could rest easy, knowing that we can explain how and why our universe came to be dominated by matter over antimatter.

But that just isn't true. Although each of Sakharov's conditions is satisfied to some degree, these known processes and events can generate only a very mild victory of matter over antimatter in the early universe—far too mild to solve the puzzle at hand. If what we know today about our universe were the entire story, then almost all of the matter would have been destroyed as our universe cooled, leaving little to form stars or planets, and providing next to no chance for the emergence of life. Our universe would be nothing like the world we see all around us.

The fact is, we just don't know how our universe came to contain so much matter and so little antimatter. The laws of physics as we currently understand them are not able to resolve this puzzle. But this realization teaches us something important about the earliest moments of our universe's history—namely, that they were very different from anything we know or understand. Unknown interactions took place, involving forms of matter and energy we have never observed in any experiment. And dramatic events unfolded, transforming the nature of our entire universe. Though this realization might raise more questions than it offers answers, this to me is precisely its allure. There is great joy to be found in the mystery of the unknown.

7

Hearts of Darkness

Science progresses best when observations force us to alter our
preconceptions.

—VERA RUBIN

WHEN YOU LOOK UP and gaze upon the night sky, the pan-
orama that you see before you consists of thousands of stars.
They dominate our view of the heavens, as well as our collective
imagination of the world beyond our own Solar System. In
many respects, however, stars don't play a particularly important role in our universe. Although bright and easily seen, they
make up only a tiny fraction—about 1.6 percent—of all of the
matter in our universe today.

Among all of the atoms in our universe, relatively few are
found within stars—or in planets, comets, asteroids, or any
other such objects. The majority of atoms instead reside among
the vast quantities of gas, mostly hydrogen and helium, which
populate so much of interstellar space. But even when we take
into account all of the gas, stars, and planets, we find that all of

the atoms together make up only a modest fraction—about 16 percent—of all matter. The overwhelming majority of our universe's material is not made up of atoms, or of any other known substance. It is invisible, or at least nearly so, and its nature is among the greatest mysteries in all of cosmology. For lack of a better name, we refer to this hidden but ever-present substance simply as dark matter.

Although we can't see the dark matter directly, we can tell that it's there from the pull of its gravity. Even if the Sun were entirely invisible, we could figure out where it was and measure its mass by studying the motion of the planets in orbit around it. Much as the Earth orbits the Sun, the Sun is in a long and slow trajectory around the center of the Milky Way. Over the next 250 million years or so—or one galactic year—our Solar System will complete an entire orbit around the Milky Way, returning to approximately the same place that it is in now. Measured in galactic years, our earliest hominid ancestors appeared only around two weeks ago, our sixteen-year-old Solar System is about to get its driver's license, and our fifty-five-year-old universe may be just beginning to think about how it may want to spend its retirement.

Just as the Earth is held in its orbit around the Sun by the force of gravity, so too is our Solar System bound to the Milky Way by gravity's attraction. From the size and speed of Earth's orbit, a first-year physics student can calculate what the mass of the Sun must be in order to maintain the Earth in its current trajectory. We can do the same thing for the Sun and its orbit around the Milky Way. When we do, we find that there must be about 100 billion suns' worth of mass within the volume enclosed by the Sun's galactic orbit. Roughly half of that mass takes the form of stars and gas. The other half, we simply do not see—it consists of dark matter. If it were not for the presence of this dark matter, stars such as our own would behave very differently. Untethered from dark matter's gravity, the stars of the Milky Way would move outward, taking on much larger and slower orbits. In some cases, such stars would even break away from the Milky Way entirely, flying away forever into intergalactic space. Without dark matter to hold them together, galaxies as we know them would largely disintegrate.

We don't yet know what the dark matter is or what it is made of. We do know, however, that it was formed sometime in our universe's first seconds, and likely in the first millionth of a second after the Big Bang. If we can come to understand this substance, we will learn not only about the nature of dark matter itself, but about our universe's earliest moments. By studying dark matter, we are studying the origin of our world.

Over the past few decades, the overwhelming majority of astronomers and cosmologists have become thoroughly convinced that dark matter exists. We see its imprint almost everywhere we point our telescopes: in the motions of stars within galaxies and of galaxies themselves within the larger systems known as galaxy clusters. The effects of dark matter can even be seen in the way that the gravity of a galaxy cluster deflects light as it passes by.

But perhaps our single strongest line of evidence for the existence of dark matter is that observed in the temperature patterns of the radiation left over from the Big Bang—the cosmic microwave background. Measurements of this radiation provide us with a map of how matter was distributed in our universe only a few hundred thousand years after the Big Bang. This map tells us that our universe was very uniform in its youth, with only the smallest variations in density. Without help from dark matter, there is no way that these density variations could have grown fast enough to form the galaxies and other large structures that we find in our universe today. The forces that act among atoms cause them to resist compression—if you don't believe me, just try squeezing a balloon. But dark matter is different. Unlike atoms, it is effectively immune to such forces, allowing gravity to compress it much more quickly, and shaping it into the scaffolding that the rest of our universe's structure would later be built upon. Long before there were any galaxies, the dark matter began to gather together into enormous clouds. It was the gravity of these dark clouds that attracted and pulled together the atoms that would ultimately go on to form galaxies themselves.

From observations such as these, we have not only been able to measure how much dark matter there is in our universe, but have also learned a limited amount about its nature. For one thing, this substance, or perhaps substances, is not made up of any of the kinds of matter we have discovered at the LHC or at other particle accelerators. Perhaps one day we'll create and observe particles of dark matter with such a machine, but not yet. For now, what we do know is that—whatever they are exactly—the particles of dark matter do not appreciably interact among themselves, or with ordinary matter, other than through the force of gravity. This means that they can pass through solid objects like a ghost, and it explains why they do not emit, absorb, or reflect any measurable quantities of light.

Should we be surprised that there is so much matter in our universe that interacts so little with the other substances around it? I certainly was when I learned this for the first time. After all, I reasoned, all of the known forms of matter communicate and interact through some combination of the electromagnetic force and the strong and weak nuclear forces. It just seemed reasonable to me that all particles would conform to this

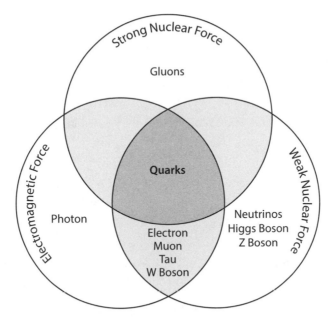

A Venn diagram classifying the known particles by the forces they either experience or communicate. All of the known particles interact through at least one of these forces, while some are subject to only two or three.

pattern. But in this line of reasoning, I had fallen for one of the oldest logical fallacies in the book.

Imagine making a list of everything in a room and then dividing the entries on that list into categories based on whether each object can be seen, heard, or felt. Many things fall within the overlap of these three criteria: my records, for example, can be easily seen, felt, and heard. Other things—like the air in the room—can be felt, but not necessarily seen or heard. So I find myself with a list of things that can be seen, heard, or felt, or some combination thereof. But no entry on my list is unseen, unheard, and unfelt.

By analogy, we can construct a list of the kinds of particles that exist in our universe, and break them down into categories based on whether they experience the electromagnetic, strong or weak nuclear forces. Of the particles contained within the Standard Model most, but not all, are subject to the effects of the electromagnetic force. The electron, muon, tau, W, and all six quarks each carry electric charge and thus interact in this way, while the three neutrinos, gluons, Z, and Higgs boson do not. Similarly, only the quarks and gluons carry "color" and therefore feel the effects of the strong force. Across all the particles in the Standard Model, only the quarks experience all three of these forces, and every particle interacts through at least one. Even neutrinos—the most feebly interacting member of the Standard Model—are subject to the weak force.

The mistake I made when I first learned about dark matter is that I didn't take into account *how* we came to discover the particles that make up the Standard Model. Just because my list of objects in the room does not include anything that has not been seen, heard, or felt does not mean that no such things are present—only that I have not become aware of them. Similarly, if there exists a particle species that does not interact through any of these three forces, we have to ask ourselves: How would it affect our experiments? How would we know whether it was there? Or that it exists at all? The answer to these questions is that, in all likelihood, we wouldn't see it or its effects in any experiment or in any laboratory. The only place we might notice it at all would be in the cosmos, where we might hope to detect it through the effects of its gravity. In other words, it would be just like dark matter. It would be dark matter.

Although the overwhelming majority of experts agree that dark matter almost certainly exists, there is not yet a complete scientific consensus on this question. For a small fraction of the astronomers and physicists who have studied this problem, the evidence for dark matter has not proven entirely persuasive. To these scientists, dark matter is not the only way that the motions of stars around galaxies could be explained—or even the patterns of galaxies and galaxy clusters found throughout our universe. Rather than postulating one or more new forms of nonluminous matter, they speculate that we may instead have misunderstood the effects of gravity.

Since the 1980s, physicists have been trying to alter the laws of gravity and motion in such a way that could explain the observed dynamics of galaxies without the existence of dark matter. This set of ideas generally falls under the umbrella of what is known as modified Newtonian dynamics—or MOND, for short. In some ways, MOND is fairly easy to implement. If you take Newton's second law of motion—force equals mass times acceleration—and change it to force equals mass times acceleration *squared*, you can more or less explain the observed behavior of stars and gas in galaxies. Of course, there's a good reason why Newton originally wrote this law the way that he did. Any number of simple experiments will confirm that force does in fact equal mass times one power of acceleration, and not two. To address this, proponents of MOND propose that this modification comes into effect only in those cases in which the acceleration being experienced is very small. Roughly speaking, this means that gravity works ordinarily here on Earth and in our Solar System, but very differently in the very low-acceleration environments experienced by stars throughout the Milky Way and in other galaxies. In these cases, MOND postulates that the force of gravity is effectively stronger than Newton—or Einstein, for that matter—had thought. And it is this strengthening of gravity, according to MOND, that holds galaxies together and creates the impression that dark matter must be present. In a universe ruled by the laws of MOND, dark matter is merely an illusion.

A moment ago, I said that MOND is fairly easy to implement *in some ways*. In others, it certainly is not. In its original formulation, MOND was presented as a direct modification of Newton's second law of motion. But this ignored everything we knew about general relativity. Given that Einstein's theory had by this time been beautifully confirmed by any number of high-precision tests, this was a major problem. Before MOND could ever be taken as a serious alternative to dark matter, it would have to be shown that it could somehow be built into a theory that is consistent with the successful predictions of general relativity.

Many physicists have tried to accomplish this over the years. Some, for example, have proposed theories in which the geometry of space and time impacts the motion of matter differently and separately from its connection with gravity. In these so-called bimetric theories, there are effectively two different values for the distance between any two points in space. To some physicists, this seemed like a promising way to explain why general relativity describes the Solar System so well by itself, while the universe on larger scales seems to require the presence of dark matter. But bimetric theories generally allow for objects to move through space at speeds faster than the speed of light, leading to major problems and even logical inconsistencies. And while it's true that such theories can sometimes be further modified in order to avoid these issues, these fixes often ruin the ability of these theories to make viable predictions for the motion of objects within our Solar System. To make things worse, these early theories of MOND made clearly incorrect predictions for how gravity should deflect light and could not explain the dynamics of very large systems, such as galaxy clusters. Although attractive to some as an abstract idea, MOND suffered from a myriad of problems, both observational and theoretical.

While problems such as these deterred many physicists from investigating MOND further, some scientists continued to search for a viable modification of gravity that could do away with the need for dark matter. These efforts looked particularly promising in 2004, when the Israeli physicist Jacob Bekenstein proposed a version of MOND that seemed to be capable of addressing many of these issues. Although Bekenstein's theory was in many ways much more complicated than earlier versions of MOND, this complexity also made it more flexible, allowing it to potentially pass all Solar System tests while also correctly predicting the gravitational deflection of light. At the time, many cosmologists were intrigued and excited by Bekenstein's proposal. Even those of us actively working on dark matter were asking ourselves whether perhaps it was all an illusion after all.

Bekenstein's theory had its day in the sun, but soon enough the dusk came. In 2006, a group of astronomers presented an observation that, for most cosmologists, convincingly demonstrated that it was dark matter and not MOND that drives the dynamics of our universe. These astronomers had been studying a system known as the "Bullet Cluster," which is actually a pair of galaxy clusters, located about 3.7 billion light years away from us. Although astronomers make measurements of galaxy clusters all the time, these clusters were different in that around 150 million years ago, they collided and passed through each other. This collision created an enormous shock wave that heated the clusters' gas to its current temperature of around 200 million degrees and produced the bullet-shaped feature that has earned this system its name.

The 2006 paper on the Bullet Cluster compared two very different kinds of measurements. First, the astronomers presented a map that described how these clusters were deflecting light as it passed by or through them. This map basically tells us how the total mass—in both visible and dark matter—is distributed throughout this system. Second, they compared this to a map of the X-rays emitted from the hot gas as measured by the satellite-based Chandra Telescope. It is this hot gas that makes up most of the visible mass in these clusters. The key finding of this paper was instantly apparent to anyone who compared these two maps: whereas the bulk of the visible matter in this system—the hot gas—was confined largely to the innermost volume of these clusters, the center of gravity was not. For the first time, astronomers had found a system in which the source of the gravity—and thus dark matter—was not centered around the same place as the visible matter. No MOND theory predicted anything like this. Knowing this, the authors of this paper gave it the title "A Direct Empirical Proof of the Existence of Dark Matter."

These days, MOND continues to face many serious problems. No version of MOND—including Bekenstein's—has been proposed that can explain the detailed features of the cosmic microwave background or the dynamics of galaxy clusters. And these problems have become only more serious as measurements have improved over the past years and decades. Although a small minority of cosmologists continue to study MOND, it has become a fringe topic—usually relegated to the outskirts of scientific discourse. But until we directly observe dark matter particles and identify their underlying nature once and for all, it seems likely that research into MOND will persist—even if only as a remote possibility.

*　*　*

Over the course of my career, a large proportion of my own research has focused on the problem of dark matter. I've offered several hypotheses for what this substance might consist of, and I've written dozens of papers discussing techniques and strategies that could potentially be used to detect and identify it. But the identity of this mysterious material is not the only question about dark matter that I desperately hope to see answered one day. Perhaps even more interesting than the particle nature of dark matter is the question of how this substance came to be formed in the early universe. By studying the nature of dark matter, we ultimately hope to learn about our universe's first moments. One day, I expect dark matter to provide us with a window into the Big Bang.

We don't know much for certain about the origins of dark matter, but we do know some things. Observations of the cosmic microwave background and the distribution of galaxies and galaxy clusters each tell us that the dark matter is nearly as old as our universe itself—within 100,000 years or so after the Big Bang, it was already in place. And although we don't know exactly how it was formed, almost any event or process that could have created dark matter would have also destroyed many of the nuclei in our universe, significantly altering the abundances of the various nuclear species that are found today. Given how successful the predictions of Big Bang nucleosynthesis have been, we have little choice but to conclude that the dark matter must have been formed well before the first nuclei—within the first few seconds after the Big Bang.

In a graduate course that I sometimes teach at the University of Chicago, I like to assign the following homework problem. I describe to the students a hypothetical form of dark matter, specifying things like the mass of its particles and how these particles interact with other forms of matter and energy. I then ask them to calculate how many of these dark matter particles will be created in the Big Bang and, more importantly, how many will survive and exist in our universe today.

The answer to this question depends, of course, on the properties with which I've endowed the hypothetical dark matter particles—their mass and the ways in which they interact. In most cases, the creation of the particles is the easy part of the calculation. Any kind of particle that experiences anything greater than the feeblest of interactions will be created in vast quantities in the early universe. During these initial stages of cosmic history, pairs of such dark matter particles would have often been created or destroyed in the collisions of other particles. In working their homework problem, the students find that regardless of how much dark matter the universe started with, the abundance evolves quickly—either increasing or decreasing until it reaches a state in which the dark matter particles are created at the same rate that they are destroyed. In other words, they are pulled toward a natural state of equilibrium with the surrounding plasma of quarks, gluons, and other forms of energy.

But this equilibrium does not last forever. As time goes on and the universe continues to expand, the temperature eventually drops to a point at which there is no longer enough energy present to create new particles of dark matter. Without any way to make more dark matter, the abundance of these particles rapidly decreases until almost all of it has disappeared from our universe. The fraction that does survive this drop in temperature depends on how the particles interact. Since the particles in question are stable, they cannot simply disappear on their own. Instead, they can be destroyed only when two dark matter particles collide and annihilate each other. The more readily this occurs, the fewer dark matter particles survive the heat of the Big Bang. In this sense, the quantity of dark matter that we find in our universe today provides us with a hint about how it is likely to interact, both with itself and with other forms of matter and energy.

To make it easier to understand how the process described in this homework problem may have played out in the early universe, picture a simple game. There is a game board with a large number of pieces randomly scattered across it. In each turn, each of the pieces is moved in a randomly determined direction, and if it touches another piece, both are removed from the game. At the end of each turn, the game board is stretched in all directions, increasing in area by 20 percent and pulling each of the remaining pieces away from each other.

In this game, the pieces are analogous to the particles of dark matter, which annihilate each other in pairs when they interact. And the game board is space itself, which is expanding rapidly during this early period of our universe's history. For concreteness, let's imagine that there are enough pieces—or particles—on the board such that 90 percent of them are removed at the end of the first turn. In the second turn, a much smaller fraction of the remaining pieces is removed. This is true for two reasons. There are ten times fewer pieces on the board, so the chance that any given piece will touch another is reduced by a factor of 10. And because the board has become bigger, this further reduces the chance that any two pieces will come into contact. At the end of the second turn, about 9.25 percent of the original pieces remain on the board, and this drops to about 7.25 percent by the end of the tenth turn. After fifty turns, just over 6.8 percent of the original pieces remain. The interesting thing is that around this time, the number of pieces on the board essentially stops changing. By this point in the game, the board is so big and the number of remaining pieces so small that you would not expect a given piece to touch another ever again, even if you played the game for an infinite number of additional turns. The number of pieces that remain on the board has simply stopped evolving. In much the same way, the abundance of dark matter in our universe may have been established through a similar process in the early universe, when it ceased to interact with both itself and with other forms of matter and energy.

The calculation involved in this homework problem hasn't changed much since I first encountered it as a graduate student some seventeen years ago or so. What it shows us—either then or now—is that in order for a stable particle species to emerge from the Big Bang with an abundance that matches the density of dark matter measured in our universe today, pairs of those particles must interact and annihilate each other through a force with a strength similar to the weak nuclear force. A stronger force than this would have caused too many of these particles to be destroyed, depleting their abundance below that needed to account for dark matter, while a force much feebler than the weak force would have allowed too many of these particles to survive. As in the story of Goldilocks, the strength of the weak force seems to be "just right" to explain how the dark matter came to be formed in the heat of the Big Bang.

The moral of this calculation, at least as it was presented to me in the early 2000s, was that dark matter—whatever it is—is likely to be made up of particles that interact predominantly through the weak force. We called this class of dark matter candidates weakly interacting massive particles, or WIMPs. Although many kinds of WIMPs had been proposed over the years, none had been directly detected in any experiment. The closest things to a WIMP that we had ever observed were the neutrinos, but these particles are far too light and fast moving to constitute much of our universe's dark matter.

Despite the fact that WIMPs had so far eluded our experiments, many of us found the logic behind this calculation to be compelling. Although no one was certain, many of us thought that it was at least likely that dark matter was made up of WIMPs. The argument behind this conclusion even had a name. We called it the "WIMP miracle."

But that was then, and this is now.

Scientists like to think of themselves as rational and intellectually honest people. And for the most part we generally are, or at least we try to be. But like everyone else, scientists too can be subject to certain human frailties. One of the most common ways in which people delude themselves is in choosing to believe something not on the strength of the evidence, but because we would like it to be true.

Before you reflexively deny that your brain falls for these kinds of logical traps, just think about that time you disregarded the poll that put your candidate behind in the upcoming election, characterizing it as an unreliable outlier. Or that other time, when you uncritically accepted as true the news story that your favorite food is somehow good for your health. If the poll had instead put your candidate ahead, you would have been much more likely to have accepted it as valid without a second thought. And if someone ever tells me that coffee is bad for me, well that's just crackpot science. No one consciously decides to think irrationally. But subconsciously, even the most rational of us do this kind of thing from time to time.

Over the past few years, I have begun to suspect that I—along with many of my scientific colleagues—may have fallen into this kind of wishful thinking when it comes to the nature of dark matter. Although the arguments behind the WIMP miracle are essentially valid, they are subject to some important caveats. If something caused the early universe to have expanded differently than we currently think it did, for example, then the results of this calculation could come out very differently. Alternatively, if the dark matter is extremely feebly interacting—even more so than WIMPs—then these particles may not have interacted often enough to have ever reached equilibrium in the first place, causing this entire line of reasoning to fall apart. Although the logic behind the WIMP miracle is indeed suggestive, it is far from bulletproof.

Complicating this further is the psychological and sociological reality that many cosmologists liked what the WIMP miracle seemed to be telling them about the nature of dark matter. More specifically, physicists realized that if the dark matter is made up of WIMPs, it would likely be possible to build experiments that, for the first time, would be able to directly detect and measure individual particles of this substance. This is something we certainly wanted to be true. After all, such an accomplishment would not only constitute the discovery of an entirely new form of matter, but would allow us to begin to measure the properties of these particles. Such experiments—it was both thought and hoped—would usher in a new era in cosmology and particle physics.

With these motivations in mind, a small army of experimental physicists set out to build the kinds of experiments that they thought would finally be able to detect individual particles of dark matter. These experiments started out small, deploying targets consisting of only a few kilograms of crystalline materials such as germanium, silicon, calcium tungstate, or sodium iodide. As time went on, however, the size and sophistication of these experiments steadily increased, relentlessly pushing the limits of their respective technologies.

When an individual WIMP collides with an atom, it behaves like a tiny billiard ball—a portion of its momentum gets transferred to the atom, giving it a sudden push. Such a push can cause the atom's nucleus to lose some of its electrons, and the moving nucleus can go on to strike a number of other atoms, leading to a chain reaction. When all is said and done, such a collision can result in a brief flash of light, heat, and even electric charge—all of which can potentially be measured by a carefully constructed detector. Fortunately for such experiments, we are being constantly bombarded by dark matter particles. If they interact even occasionally with ordinary matter—as our calculations suggest WIMPs should—then these devices should be able to detect these collisions.

But in many ways, the detection is the easy part. The real challenge lies in the fact that dark matter particles are far from the only thing that can interact with these kinds of detectors. Although dark matter is constantly passing through all of us, we are also immersed in a violent sea of ordinary atoms and radiation. In order to isolate and identify the occasional push from a dark matter particle, one must first quiet the surrounding sea of atoms. The universe, after all, is a very noisy and volatile place.

To this end, experimental physicists set out to build ultrasensitive detectors, carefully shielding them from the ambient background and deploying them in deep underground laboratories where they are protected from the most distracting forms of cosmic radiation. Although these detectors are small in comparison to those such as the LHC, they are in some ways no less sophisticated. And over the past fifteen years or so, their sensitivity has been increasing at a staggering rate. Compared to modern dark matter detectors, those from a decade ago seem almost quaint. The state of the art in this field today makes use of enormous detectors, consisting of multiple tons of liquid xenon. Xenon has at least two major advantages as a dark matter target. As a noble element (those in the far right column of the periodic table), it's very stable and highly nonreactive, both of which help to minimize the kinds of activity that might compete with a signal from dark matter. And second, its large size—consisting of between 124 and 136 total protons and neutrons—makes it a big and effective target for incoming particles of dark matter. Over the past two decades, the power of detectors such as these have been increasing at an exponential rate—more than doubling their sensitivity every year or so. This makes the growth in computing speed experienced over the past half century—Moore's law—seem sluggish in comparison.

In the early 2000s, many cosmologists and particle physicists—including myself—thought that it was likely that we would successfully detect particles of dark matter over the next decade or so. I wrote at the time that I would take an even-odds bet that such a discovery would take place between roughly 2008 and 2015. And although that span of time is now behind us, no such discovery has taken place.

Experiments such as these present their results in terms of a quantity known as the dark matter's scattering cross section with protons and neutrons. Roughly speaking, you can think of this cross section as the size of a target proton or neutron as experienced by the dark matter—if a dark matter particle passes through this area, it is likely to strike the target. If the particles that make up the dark matter are about 100 times heavier than a proton, for example, the leading experiments tell us that its cross section must be no larger than about 10^{-46}

square centimeters. For comparison, the cross section for low-energy photons to interact with an electron through the electromagnetic force is about 7×10^{-25} square centimeters—more than a trillion-billion times larger than those being tested by modern dark matter detectors. Even if the dark matter were subject to the kinds of interactions experienced by neutrinos, this would correspond to a cross section roughly a million times larger than those we have already put to the test. Whatever the dark matter is, it is far more elusive than any form of matter we have ever seen.

Just to be clear, the failure to detect particles of dark matter has not been for a lack of experimental progress. From a technological standpoint, the experimental search for dark matter has been incredibly successful. Modern experiments are roughly 10,000 times more powerful than those that were operating in 2006—a spectacular feat. And yet no signals appeared in these experiments. No unexplained flashes of light, heat, or charge. Nothing but silence. But even silence is informative. These experiments may not have discovered the identity of dark matter, but they have taught us a great deal. We now know that the dark matter cannot be as we once imagined it.

Our failure to detect dark matter in underground experiments and at the LHC has had a palpable effect on the scientific community. Although it remains the case that a discovery could still plausibly lie right around the corner, most of the scientists studying dark matter today will acknowledge that many of their favorite dark matter candidates should have been detected by now. The hunt for dark matter was never supposed to be easy. But frankly, we didn't expect it to be this hard.

In the absence of dark matter's discovery, the scientific community has begun to redirect its efforts toward new, and sometimes very different, ideas. Although this state of affairs can certainly be frustrating at times, it has also had the positive effect of ushering in an explosion of theoretical work related to dark matter. For example, theories in which the dark matter is but one of several particle species that make up what is known as a hidden sector have recently attracted a great deal of attention. Because hidden sector particles do not directly interact with ordinary particles, they can be very difficult to detect in underground experiments and hard to produce at accelerators. And alongside such theoretical advances, we are also making considerable experimental progress. Scientists are developing and carrying out new and very different kinds of experiments, with the goal of testing an increasingly wide range of previously overlooked dark matter candidates.

The remarkable progress of underground experiments has thrown the field of dark matter into a state of major disruption. It turns out that the dark matter is probably very different from what most of us had once thought. The stubborn elusiveness of dark matter has left myself and many of my colleagues both surprised and confused. We had what seemed like very good reasons to expect particles of dark matter to be discovered by now. And yet the hunt continues, and the mystery deepens. In droves, physicists are returning to their chalkboards, revisiting and revising their assumptions. With bruised egos and a bit more humility, we are desperately attempting to find a new way to make sense of our world.

8

A Beacon in the Dark?

When I see a bird that walks like a duck and swims like a duck and
quacks like a duck, I call that bird a duck.

—ATTRIBUTED TO JAMES WHITCOMB RILEY

A WHILE BACK, a friend of mine told me that he was having a
hard time visualizing what it was like to do my job. He knew
that I was a physicist, and that I worked on topics like dark
matter and cosmology. But what did it mean for someone to
"work on dark matter?" He put it to me this way: "So you walk
into your office. You take off your coat. You get yourself a cup
of coffee. How do you know what to do next?"

This question has motivated me to tell you about some of my
own research in this chapter. I'm going to start this story back
in 2009, when I discovered a strange gamma-ray signal from the
center of our galaxy. It's possible—and some would say likely—
that these gamma rays are generated by a collection of thousands of rapidly spinning neutron stars, called pulsars. By most
standards, this in itself would be an exciting discovery. But even

more exciting is the possibility that these gamma rays are not related to pulsars, but are instead being produced by particles of dark matter. If this is indeed the case, then this signal would provide us with our first real insight into the nature of this mysterious substance, as well as an important glimpse into how the dark matter was formed during the first millionth of a second after the Big Bang.

By the end of this chapter, I'll have taken this story up to where things stand now. But even then, I won't be able to tell you how it ends. The last chapter of this tale has not yet been written.

One morning in late August of 2009, I was in my office on the sixth floor of Fermilab's main high-rise tower, Wilson Hall. I had already taken off my coat and had gotten myself a cup of strong coffee. Now I was sitting down with my feet on my desk and my eyes fixed on the screen of my laptop. Over the past few weeks, I had been writing a computer program to analyze the data that had been collected by a gamma-ray telescope known as Fermi.[1]

Fermi is not like what most people think of when they picture a telescope. It has no lenses and no mirrors, and is blind to the kinds of light that our human eyes can see. Instead, it is designed to perceive the most energetic forms of light, gamma rays, which are hundreds of millions of times more energetic than visible photons. And whereas many modern telescopes are positioned on the tops of mountains, the Fermi Telescope is located in an even more remote location—onboard a satellite 340 miles above the surface of the Earth.

---

1. The Fermi Telescope has no direct relation to Fermilab, other than the fact that both were named after Enrico Fermi, who made important contributions to both particle physics and astrophysics.

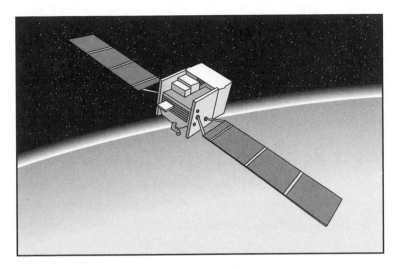

An artist's depiction of the Fermi Gamma-Ray Telescope.

On June 11, 2008, Fermi was successfully launched from Cape Canaveral aboard a Delta II 7920-H rocket into low Earth orbit. In the years since, it has been steadily sweeping across the sky, recording the energy and direction of the gamma rays that strike its carefully designed detectors. Fermi was designed to produce the most detailed map of the gamma-ray sky that has ever been seen and to study in detail many of our universe's most volatile objects and environments. For example, supermassive black holes in the act of swallowing large quantities of matter can produce bright jets of gamma rays, visible to Fermi over distances nearly as large as the observable universe itself. To date, Fermi has detected gamma rays from well over a thousand of these objects. In addition, exotic exploding stars known as gamma-ray bursts were also a top target of Fermi's instruments. And within the Milky Way, Fermi was expected to map out and study our galaxy's population of rapidly spinning neutron stars, called pulsars. The scientists and engineers who designed and built Fermi had every reason to expect that their telescope would reveal a great deal to us about the most extreme objects in our universe. But in addition to studying these kinds of objects, there was also a chance that Fermi could accomplish something even more spectacular. If we were lucky, Fermi might reveal to us the nature of dark matter.

The idea that gamma-ray telescopes could be used to study dark matter is not a new one. On Valentine's Day of 1978, more than thirty years before Fermi's launch, two incredibly prescient papers were published side by side in the journal *Physical Review Letters*. In these articles, two independent groups of scientists—consisting of Jim Gunn, Ben Lee, Ian Lerche, Dave Schramm, Gary Steigman, and Floyd Stecker—argued that a small but important fraction of the dark matter particles in our universe today is likely to be undergoing the process of annihilation. Although dark matter particles emit negligible amounts of light in isolation, it is entirely plausible that they produce a far brighter signal when concentrated together in groups. In particular, when pairs of dark matter particles come into contact with and annihilate each other, their mass can be converted into other forms of energy, including that of gamma rays. If we could build telescopes capable of detecting those gamma rays, we could learn not only about the nature of dark matter itself, but also map out its distribution throughout our universe—dark matter cartography. The information provided by such an observation could even be used to infer how the dark matter was generated in the first moments after the Big Bang. It was a very exciting prospect, and although I was only one year old at the time, I like to think that it was on that Valentine's Day that I—always a hopeless romantic—became destined to one day search the gamma-ray sky for signs of dark matter.

*   *   *

I wrote my first paper about using gamma rays to study dark matter back in 2002, while I was still a graduate student. Over the years that followed, I continued to think about this problem, working to pin down what a gamma-ray signal from dark matter might look like and figure out ways to distinguish this kind of signal from the energetic photons produced by more ordinary astrophysical sources. We didn't expect it to be easy, but Fermi represented a huge leap forward in technology, and it didn't seem ludicrous to think it could discern a signal of dark matter from among the backgrounds.

Of all of the places to look, the most promising seemed to be the center of our own galaxy, the Milky Way. Our best measurements and computer simulations indicate that the density of dark matter is quite high near our galaxy's center, which means that we should expect the dark matter particles present there to be interacting with and annihilating each other at a high rate—generating large numbers of gamma rays in the process. Although there were many unknowns involved in this calculation, we generally expected that these dark matter interactions would produce a bright cloud-like signal of gamma rays, centered on our galaxy's own supermassive black hole and extending outward in every direction.

So let's return to that morning in August of 2009. Fermi had been in orbit for over a year by then, but it had been just a few weeks since its first batch of data had become available to outside scientists like me. After a few weeks of work, the analysis program I had been writing was finally coming together. If it worked the way it was supposed to, this program would take the observed distribution of gamma rays—both in terms of direction and energy—and use this information to deduce whether these energetic photons were coming from dark matter, astrophysical processes, or some combination of the two. For the first time, I executed the code and plotted some of the results. When the results appeared on my screen, I didn't know what to think.

As Isaac Asimov famously said, it's not words like "eureka" that herald scientific discovery, but rather "that's funny . . ." I could feel my face take on a look of consternation as I leaned toward my computer and tried to figure out what it was that I was looking at. The data that I had plotted certainly didn't look at all like what I had been expecting. I had anticipated seeing a smooth spectral shape that could easily be explained by rather ordinary astrophysical processes, like cosmic rays interacting with gas and radiation. From a spectrum like that, I could have calculated and published a new and more powerful upper limit on the dark matter annihilation rate, which would have helped us to limit the range of possible theories that could possibly describe the dark matter of our universe. It would have been a nice paper. But even at first glance, this data looked different. To my surprise, the spectrum of the gamma rays from the direction of the Milky Way's center was not particularly smooth, but instead exhibited a strange bump-like feature across a range of energies centered at around two gigaelectron-volts. I leaned forward at my desk and stared at the screen. Disoriented, I kept thinking to myself that this was not what we had expected to see. Instead, this really looked a lot like dark matter.

I showed these early plots to some of my colleagues, including Lisa Goodenough, who was my collaborator on the project and a graduate student at New York University at the time. Neither of us was sure what to do. A paper presenting a new upper limit on the dark matter annihilation rate would have been uncontroversial and relatively easy to write. But the data before us wasn't calling for that kind of paper. So after another month and a half of testing and examining our analysis, we started writing.

Our paper was titled "Possible Evidence for Dark Matter Annihilation in the Inner Milky Way from the Fermi Gamma Ray Space Telescope."

It's fair to say that many of our colleagues did not respond well to our paper. Publicly, some of these scientists objected that we had employed an overly simple model to describe the gamma-ray backgrounds, potentially skewing our results. Others worried that we had not adequately accounted for some of Fermi's instrumental effects. On both of these points there was some truth. Our analysis of this data set had been fairly rudimentary, and there was good reason to be skeptical. More concerning still, some of the scientists associated with the Fermi Telescope told us that they had looked at the same data themselves and could not reproduce our results. As far as they were concerned, there was nothing like a dark matter signal coming from the center of the Milky Way.

After six months or so of largely negative press and critical comments about our work, Lisa and I returned to the data. Frankly, we weren't sure what to think. Although the criticism seemed to include some valid points, many of the loudest objections seemed way off base. The one thing we knew for certain was that the Galactic Center is an interesting place to observe with a telescope like Fermi, and so far no one had really established what we could learn from it. At the same time, we knew that to convince our colleagues that they could trust our conclusions, we would have to dig deeper. So in our second analysis, we more carefully took into account factors such as the resolution of the instrument and modeled in more detail many of the astrophysical sources that constituted the primary backgrounds. Yet once again, when the analysis program was run, a strange bump appeared in the gamma-ray spectrum. And this time, we could more clearly make out the pattern of the gamma-ray signal on the sky. In each of these respects, it continued to look a lot like dark matter.

We published our results in a second paper, and although the reception was mixed, it did a lot to persuade me personally that this signal was real and that it might very well be the consequence of dark matter. I gave dozens of seminars and conference talks on this study, trying to persuade as many of my colleagues as I could that this was an interesting result and that it should be taken seriously. In parallel, I continued to study the data in new ways, publishing a third analysis in 2011 with Tim Linden, who at the time was a graduate student at the University of California, Santa Cruz.

Starting in 2012, other groups of scientists began to present the results of their own analyses of this data. In each case, these groups identified a signal that was very similar to what I had reported with my collaborators. I felt both vindicated and excited and was eager for the debate to finally move on from the question of whether this signal exists to that of what generates these gamma rays. But this shift didn't happen the way that I had imagined it would. Even after these new studies were presented, many of the scientists affiliated with the Fermi Telescope continued to insist that there was no signal there at all.

During this time, a number of different groups of scientists were working hard to better understand this collection of data. Throughout 2013, I worked as part of a group of scientists at Harvard, the Massachusetts Institute of Technology, Fermilab, and the University of Chicago who were using a variety of different analysis techniques and approaches to test and scrutinize the data from Fermi. In February of 2014, we presented our results in a long paper that, at long last, persuaded many scientists that the signal is in fact real. Even NASA put out a press release on behalf of the scientists operating Fermi, acknowledging that this signal was truly present in the data. It had taken almost five years, but a consensus had begun to emerge that this signal really existed. Whether or not it was a consequence of dark matter, however, was another question altogether.

By this time, we knew a lot about the gamma-ray signal Fermi had been observing from the Galactic Center. Its intensity, its spectral shape, and even its extent across the sky had all been reasonably well measured. And in each of these respects, the measurements were remarkably consistent with annihilating dark matter. For example, if the signal is indeed coming from dark matter, then the way that these gamma rays are distributed across the sky should correspond to the dark matter's spatial distribution throughout the inner volume of the Milky Way. Astrophysicists specializing in the formation and development of galaxies had previously made predictions for how the dark matter should be distributed within such systems, and when we compared these two things—the data and the theory—they were a good match. The pattern of these gamma rays across the sky looks a lot like what many of us had expected from dark matter.

In other ways, we could use the Fermi data not only to determine whether dark matter is responsible for this signal, but also to measure the properties and characteristics of the dark matter particles themselves. In particular, the spectrum of the gamma rays that are produced in these kinds of interactions tells us a lot about dark matter particles and how they annihilate each other. In the case of the Fermi data, the measured spectrum provides a good match to what we would predict from dark matter particles that are roughly fifty or sixty times as heavy as a proton and that produce mostly quarks or gluons when they are annihilated. Furthermore, the overall intensity of the signal tells us that these particles must be annihilating each other at a rate similar to what we would expect from interactions through the weak force. In other words, the measured intensity seems to match what we would have predicted from arguments based on the WIMP miracle. So in yet another way, these gamma rays looked a lot like a signal of dark matter.

But even in light of such considerations, the scientific community is intrinsically skeptical and tends to be reluctant to accept new claims of discovery. In most cases, this is a good thing. After all, the overwhelming majority of claimed discoveries turn out to be false alarms, and if you invested any significant amount of your time on every such claim, you would never get anything else done. And to be fair, although the characteristics of this gamma-ray signal do broadly match what was expected from annihilating particles of dark matter, this does not prove that dark matter is the correct interpretation. To put it another way, just because something quacks like a duck does not mean that it is necessarily a duck.

From as early as 2010, a number of scientists—including myself—had been actively thinking about various kinds of astrophysical sources or mechanisms that might be able to produce a signal like the one we had discovered in the Fermi data. As the quantity of the data grew and the quality of the analysis techniques improved, however, most of these proposals were gradually ruled out. For example, some of us speculated early on that most of the gamma rays that make up this signal might originate from the Milky Way's supermassive black hole. Astronomers have known for some time that an enormous black hole resides at the center of our galaxy, about 4 million times as massive as the Sun. It is certainly plausible that such an object could accelerate particles and create high-energy radiation of the kind that Fermi had observed. But today we know that the gamma-ray signal in question extends at least 10 degrees or more in all directions away from the Galactic Center—much too wide to originate from a single, central source (for comparison, the Moon and Sun are each about half a degree in diameter, as seen from the Earth). We also considered the possibility that a recent series of events near the Galactic Center may have accelerated a huge number of electrons or protons to incredibly high speeds, resulting in the production of gamma rays as these particles traveled outward and interacted with the surrounding gas and starlight. Again, however, scenarios such as these became increasingly problematic as our understanding of the data improved. Such cosmic-ray outbursts now seem unlikely to have had much to do with this signal.

There is, however, one other explanation for this signal that continues to appear quite promising. If you were to survey a group of gamma-ray astronomers about what they think is most likely to generate the gamma rays in question, you would almost certainly hear the following word: pulsars.

Throughout the course of their lives, stars exist in a balance between the force of gravity pushing inward and the pressure of nuclear fusion pushing out. For most stars, this state of equilibrium persists for billions or even tens of billions of years. Ultimately, though, the nuclear fuel has to run out. And when it does, gravity can take hold of a star and transform it beyond recognition.

When the Sun's core runs out of hydrogen some 5 billion years from now, the force of gravity will compress its vast mass into a volume the size of the Earth, forming a compact and thoroughly dead object known as a white dwarf. The larger a star is, the more rapidly it consumes its fuel—massive stars burn hot, and they die fast. Stars with tens or hundreds of times as much mass as the Sun stop undergoing fusion after only millions of years, ending their lives in a spectacular explosion animated by the intense power of gravity—a supernova. When the explosion is over, the most massive of these stars leave behind a black hole. Black holes are one of nature's most extreme objects and are the perfect representation of gravity's total victory over all other facets of nature—objects so entirely adherent to the will of gravity that even light cannot escape.

Although gravity always wins such contests in the end, it does not always win so completely. A star that starts out with between ten and thirty times as much mass as the Sun will explode as a supernova, but will not collapse entirely into a black hole. Instead, these stars ultimately form another kind of strange and exotic object, known as a neutron star.

Though perhaps not as extreme as black holes, neutron stars are mind-boggling all the same. Such stars consist of an entirely different form of matter than anything you or I have ever experienced. When the precursor to a neutron star is collapsing, the force of gravity is so relentless as to actually transform the nature of its substance itself. In this collapse, the atoms are driven together with incredible force. And although they resist this compression with the full power of the electromagnetic force, it is not enough. When all is said and done, nearly all of the electrons and protons are destroyed—utterly eradicated. In their place is a very different and unimaginably dense form of matter, consisting entirely of neutrons.

It is worth pausing for a moment to wrap your head around the incredible density of a neutron star. The electromagnetic force strives to keep electrons as far apart from each other as possible, limiting how much ordinary matter can be squeezed into a given volume. This makes it almost impossible to compress any ordinary material to a density higher than about 10 or 20 grams per cubic centimeter. Even the densest material on Earth—the metallic element osmium—has a density of only 22.6 grams per cubic centimeter. But unlike ordinary forms of matter, the material that makes up a neutron star has no electrons or protons to resist compression. As a result, there is little to stand in the way of gravity's will, and the collapsing star is driven together until more than a Sun's worth of mass is confined within a volume the size of a small city—forming a substance that is trillions of times as dense as the densest material found on Earth.

This sudden act of compression has another consequence for the collapsing star. Most stars gently rotate about their axis. The Sun, for example, completes a rotation once every twenty-five days or so. But whenever a spinning object is compressed, it spins faster. Every fan of figure skating will be familiar with this fact—just picture a skater pulling her arms in toward her body in order to accelerate the rate of her rotation. When a star collapses into a neutron star, it too begins to spin more quickly. Much more quickly. If our Sun were to be suddenly compressed into the size of a neutron star, it would be left spinning at a rate of nearly one thousand rotations per second. Young neutron stars are thus not only ridiculously dense, but are also spinning at spectacular speeds.

The sheer quantity of energy stored in the rotation of a young neutron star can be staggering. And this energy gradually escapes into the world around it, taking the form of powerful beams of radio waves and gamma rays. As a neutron star rotates, these beams point in different directions, like a lighthouse. From our vantage, the beams of a given neutron star might be pointed at us once or maybe twice per rotation, leading to alternating patterns of intense brightness and darkness. These

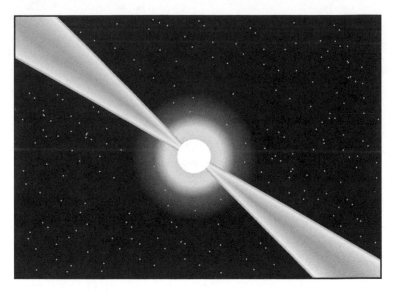

Pulsars are rapidly spinning neutron stars that emit powerful beams of radio and gamma rays.

pulses of light that characterize such neutron stars have earned these objects their names—pulsars.

Pulsars are fascinating objects in their own right. By observing these extreme stars, astronomers have learned a great deal both about this strange form of all-neutron matter and about the evolution of stars. Furthermore, the effects of gravity are so intense near these objects that observing them enables us to test the boundaries of Einstein's theory of general relativity. For these and other reasons, pulsars are the objects of affection of many astronomers. But to those of us hunting dark matter, pulsars are often the bane of our efforts. Despite all the reasons to be fascinated by these objects, there are few things in our universe that I hate more than pulsars.

*   *   *

Since the first pulsar was discovered in 1967, radio astronomers have observed a few thousand of these objects scattered throughout the Milky Way. And using the Fermi Telescope, we have detected gamma rays from over 200 of them. Among both radio and gamma-ray astronomers, few objects have been more intensely scrutinized than pulsars.

Prior to Fermi, however, we didn't know very much about how pulsars generate their gamma rays. Fermi's predecessor, the EGRET Telescope, observed gamma rays from a handful of such objects, but it wasn't until Fermi that this area of investigation really came of age. We now have detailed measurements of many pulsars, revealing how the intensity of their gamma-ray emission changes over the course of their rotations, as well as the shapes of their gamma-ray spectra. In fact, it is the spectral shape of pulsars that has been so frustrating to so many dark matter hunters—frustrating because it coincidentally resembles the signal we predict from dark matter. In the hunt for dark matter, pulsars can often act as a decoy.

Much like the signal from the center of the Milky Way observed by Fermi, the gamma-ray spectrum from most pulsars includes a distinctive bump-like feature that peaks at around 2 gigaelectron-volts. A large number of active pulsars near the center of our galaxy would appear to Fermi as a diffuse cloud of gamma rays, with a peaked spectrum. In other words, if there were a large population—tens of thousands—of gamma-ray emitting pulsars distributed in the right way throughout the center of our galaxy, it would be very difficult for a telescope such as Fermi to distinguish their gamma rays from an authentic signal of dark matter.

So which is it? Does this signal come from dark matter, or from pulsars? The fact is, we just don't know yet. On the one hand, unlike any particular dark matter particle that I might hypothesize, we actually know that pulsars exist. This simple fact is a good reason to take this possibility seriously. But on the other hand, we've identified only one pulsar near the Galactic Center so far, and it isn't the kind that we would expect to contribute to this kind of gamma-ray signal. If there really is a large population of pulsars located near the Galactic Center, we should have already seen the most exceptional of these objects stand out as bright individual sources of gamma rays—and we don't see anything like that yet.

If it turns out that pulsars are responsible for this signal, they will not be able to hide from us forever. New radio telescopes, consisting of thousands of coordinated antennas, are currently being developed and will be able to see many of the pulsars that might make up such a population. If they exist, between dozens and hundreds of these pulsars will be identified in the years ahead. But if large numbers of pulsars are not discovered by these radio surveys, that would all but rule out the possibility that the gamma-ray signal is generated by such objects and would substantially bolster the case that it originates from particles of dark matter concentrated in our galaxy's core.

As I am writing this, we still do not know what generates the signal Fermi observed. But one day—perhaps soon—this will change. What would happen if we were to definitely establish that these gamma rays come from dark matter? What would this teach us about the nature of dark matter? And what would it tell us about our universe and its early history?

If the gamma rays observed from the center of our galaxy are confirmed to be an authentic signal of dark matter, this would provide us with a huge cache of invaluable information. First, the shape of the spectrum tells us a lot about the mass of the individual dark matter particles and the kinds of matter and energy they produce when they annihilate each other. Knowing this would dramatically narrow the range of hypothetical particles that might make up this substance. And second, the overall intensity of this signal allows us to infer the strength of the interactions that cause the dark matter to be annihilated. In the case of the gamma rays observed from the Galactic Center, their brightness tells us that the dark matter is being annihilated at a rate similar to that predicted for interactions through the weak force. Or to put it another way, this signal provides us with considerable support for the WIMP miracle.

By studying how particles of dark matter annihilate each other in the center of our galaxy today, we can also learn about the earliest moments of our universe's history. The more readily that pairs of dark matter particles annihilate each other, the less of this substance survives the heat of the Big Bang. But conclusions such as these are contingent on a number of assumptions about the early universe. For example, when we calculate how much dark matter should have survived the Big Bang, we generally assume that the energy in the early universe was overwhelmingly dominated by a soup of particles traveling at or near the speed of light. From the measured abundances of the light nuclei—hydrogen, deuterium, helium, and lithium—we can be confident that rapidly moving particles made up most of the energy in our universe from a second or so after the Big Bang until about 100,000 years later, when dark matter instead began to play the leading role. But when it comes to our universe's first second or so, we currently have no way to know for certain how fast it was expanding or what kinds of matter or energy filled space. It's entirely possible that the early universe expanded either more quickly or more slowly than we usually assume.

This is one of the reasons why studying dark matter is so important. By measuring the ways in which the dark matter interacts, we can learn about our universe during these very early times. The kinds of dark matter particles that could generate a gamma-ray signal like the one observed from the Galactic Center would have been overwhelmingly formed in the first millionth of a second or so after the Big Bang. In much the way that cosmologists have learned how fast our universe was expanding in its first minutes by studying the light nuclear elements, we may be able to learn how rapidly our universe was expanding at much, much earlier times by measuring the interactions of dark matter. Today, we are blind to the first seconds of our universe's history. If the gamma-ray signal that my collaborators and I discovered is in fact generated by the interactions of dark matter, then it could provide us with a new way to study the early universe. Dark matter may very well be the window through which we get our first glimpse of the first millionth of a second after the Big Bang.

Radically Rethinking Dark Matter

It doesn't make any difference how beautiful your guess is. It does
not make any difference how smart you are, who made the guess,
or what his name is—if it disagrees with experiment it is wrong.

—RICHARD FEYNMAN

THERE ARE GOOD REASONS to be optimistic about the ongoing efforts to discover and understand the particle nature of
dark matter. For one thing, the gamma-ray signal observed from
the Galactic Center may very well be the result of dark matter
particles interacting and annihilating each other in pairs. If this
is true, then we may be on the cusp of finally identifying the
nature of this mysterious substance and learning how it was
formed in the early universe.

But what if instead, we confirm that this signal is in fact generated by thousands of pulsars or by some other collection of
astrophysical sources or mechanisms? Where would this leave
us? The absence of any signs of dark matter in underground
experiments and at the LHC has been extremely disappointing.
And as these experiments have become increasingly sensitive,

it has become only more difficult to explain why no such signals have appeared. After all, if the dark matter really does consist of WIMPs, then we have every reason to think that these experiments should be able to detect the particles that make up this substance. Yet they haven't.

This state of affairs has prompted many cosmologists to re-evaluate their ideas about dark matter and its nature. And with every new experiment that fails to detect these particles, the more we feel compelled to revisit and challenge our long-held ideas about this substance. Perhaps the dark matter does not consist of the kinds of particles that we have been hunting for so long. Instead, the dark matter may take a very different form, and may have been generated very differently in the first moments after the Big Bang.

The lack of a clear and definitive detection of dark matter is not the only thing that has shaken the confidence of many physicists in recent years. In the years leading up to the operation of the LHC, most of my colleagues—including myself—expected this machine to usher in a new era in particle physics, in which many new forms of matter and energy would be discovered. Although the scientists operating the LHC did discover the long-sought-after Higgs boson in 2012, no other new signals have since appeared. The LHC has enabled us to make incredibly powerful and precise measurements of the known particles and their interactions, but it has not yet brought forth the kinds of discoveries that many of us had hoped for—and even expected.

Any number of arguments have been made for why a machine like the LHC would likely discover new kinds of particles. The most straightforward—although not necessarily the most persuasive—is the argument based on past experience.

Throughout the twentieth century, whenever physicists built substantially bigger and more powerful accelerators, these machines almost always led to new discoveries. The previous machine to collide high-energy protons—the Fermilab Tevatron—discovered the top quark in 1995, and its European predecessor—CERN's Super Proton Synchrotron—revealed the W and Z bosons in 1984. Accelerators in the 1970s were used to discover the gluon, the tau lepton, and the charm quark. Although the physicists building and operating these machines didn't always know what they were looking for, the discoveries did come. Over decades, this lineage of experiments established an impressive record of discovering new forms of energy and matter, as well as new ways in which these particles behave and interact.

But the historical record was not the only reason why so many physicists were confident that new discoveries would be made by the LHC. Since the late 1960s, the theory known as the Standard Model has served as the foundation of particle physics. One after another, each of the seventeen particles predicted by this theory was discovered. The W and Z bosons, top quark, and Higgs boson were all predicted by the Standard Model decades before they were directly observed. And as the properties of these and other particles were measured, they were found—without exception—to precisely match the predictions derived from the mathematics of this theory. Based on its incredible predictive power and long list of empirical confirmations, the Standard Model has reasonably been called the most successful theory in the history of science. But despite all of these triumphs, the Standard Model also has a major problem. On its own, the Standard Model is a broken theory. It cannot be complete.

According to the Standard Model, interactions of particles with the Higgs boson—or more precisely, with the Higgs field, which permeates all of space—cause those particles to slow down and carry more energy. In other words, interactions with the Higgs field cause particles to acquire mass. This process works in the other direction as well, with each interaction contributing to the mass of the Higgs boson itself. And therein lies the problem. When we account for all of the contributions from these interactions, we are forced to a conclusion that is simply not compatible with the universe that we see all around us. This calculation tells us that the Higgs boson should be trillions of times heavier than it actually is.

There are several very different, but equally correct, ways that physicists think about this problem. One of these is to formulate it in terms of a comparison between gravity and the weak force. The problem lies in the fact that the force of gravity is amazingly feeble compared to the other known forces—roughly 10^{32} times weaker than the weak force. This enormous difference in the strengths of these two forces cannot be easily explained within the context of the Standard Model. Instead, our calculations say that the interactions of the Higgs boson should drive its mass—along with the masses of the Z and W bosons—dramatically upward, which in turn would make the weak force much, much feebler than it actually is. Based on what we know about the Standard Model, the dynamics of this mechanism should make the Higgs boson so heavy that we could never see it with any accelerator and the effects of the weak force so feeble as to be entirely imperceptible. And yet this is not the case. The Higgs boson has a comparatively modest mass—roughly 133 times that of a proton. And the weak force, at least when compared to gravity, is not very weak at all.

This problem—known as the hierarchy problem—has received a tremendous amount of attention from theoretical physicists, going back decades. Although many solutions have been proposed over the years, most of these only mitigate the problem without solving it entirely. There is one idea, however, that could truly tame the Higgs, stabilizing its mass and explaining how such an enormous hierarchy between the strengths of gravity and the weak force could persist. This idea is known as supersymmetry, and it has captivated the imaginations of more physicists than perhaps any other idea in all of contemporary physics.

Although supersymmetry as a mathematical construction goes back to the 1970s, it really wasn't until the early 1980s that it began to take on something similar to its modern form. Supersymmetry hypothesizes that there is a close connection between two different groups of particles—fermions and bosons. Fermions include all of the leptons and all of the quarks, while the photon, gluon, W, Z, and Higgs are each bosons. The basic idea behind supersymmetry is that each fermion is directly related to and partnered with a boson with many of the same properties, and vice versa. For example, in order for the electron to exist, there must also be a bosonic superpartner with the same electric charge and other quantum properties—the selectron. And because the photon exists, supersymmetry posits that nature must also include its fermionic counterpart—the photino.

Despite the fact that none of the superpartner particles posited by this theory have ever been directly observed in any experiment, many physicists have found the allure of supersymmetry to be too strong to resist. For one thing, the mathematical structure of supersymmetry is remarkably deep and beautiful. Though many people are surprised to learn it, aesthetic considerations have often played an important guiding role in theoretical physics. Furthermore, if supersymmetry is in fact embedded within the fabric of nature, this would make it much easier for physicists to understand how the different forces of nature might fit together within a common unified framework. String theory, for example, currently represents our most promising idea for how the quantum nature of the Standard Model might be self-consistently unified with Einstein's theory of general relativity. But it turns out that string theory could be a viable theory of quantum gravity only if nature is supersymmetric. In this and other ways, supersymmetry's existence would help us to make sense of some of nature's most perplexing aspects and enable us to build more powerful, yet simpler theories that can explain more about our universe.

In addition to aiding in these more grandiose aspirations, supersymmetry has an uncanny ability to solve the hierarchy problem. When they contribute to the mass of the Higgs boson, fermions and bosons push in opposite directions. Roughly speaking, each of the particles of the Standard Model has a number that quantifies how much it contributes to the mass of the Higgs. You can think of this as a number that is positive for each fermion and negative for each boson. To calculate the total contribution from all of these particles, you add all the numbers together. When you are done, you should expect to get a total that is very roughly equal to the size of the biggest individual contribution—the one that arises from interactions with the top quark.

But now imagine that our universe is supersymmetric. For every positive contribution from a given fermion there is automatically a corresponding negative contribution from its partnered boson. When the totals are summed, one finds that the contributions from the fermions almost exactly cancel those from the bosons, making the total mass of the Higgs much lower than we would have expected without supersymmetry. As long as the superpartner particles are not too heavy, this can easily explain why the Higgs is the way that we find it to be in our world.

Along with the hierarchy problem, many physicists thought that supersymmetry might also offer a solution to the problem of dark matter. In many supersymmetric theories, the lightest of the superpartner particles is stable and weakly interacting— precisely the features that one would want in a candidate for the dark matter of our universe. This particle might be the super-symmetric counterpart of the photon, Z, or Higgs boson, or a combination thereof. In any case, such a superpartner would constitute exactly the kind of particle that we could call a WIMP.

In the years leading up to the LHC, supersymmetry was all the rage. The most widely studied candidates for dark matter were supersymmetric particles, and a significant majority of particle physicists expected that this machine would ultimately discover at least some of the particles predicted by supersym-metry. But as time went on, no signs of the superpartners ap-peared. Supersymmetry contained such a beautiful confluence of attractive features that it was tempting to take for granted that it would be manifest in nature. And many physicists did just that. The problem is, if supersymmetry really does exist, we probably should have seen it at the LHC by now. Yet we haven't.

Because of the LHC, we now know that if they exist at all, at least some of the superpartners must be at least ten times or so as heavy as the Higgs boson, and more than a thousand times heavier than the proton. And although some physicists argue that it is not all that surprising that the superpartners could be this heavy, the fact is that most of my colleagues had not ex-pected this. As a consequence, enthusiasm for the very possibil-ity of supersymmetry has waned. Many of us are now much less certain that nature is supersymmetric and much less confident that we understand the hierarchy problem or how it might be solved.

The results of the LHC have shaken the science of particle physics to its core. Throughout the second half of the twentieth century, theoretical particle physicists have over and over again been able to successfully predict the kinds of particles that would be discovered as accelerators became increasingly powerful. It was a truly impressive run. But our prescience seems to have come to an end—the long-predicted particles associated with supersymmetry have stubbornly refused to appear. Perhaps supersymmetry or something like it is right around the corner, and our confidence will soon be restored. But right now, it's fair to say that we simply do not understand how to solve the hierarchy problem, how to complete the Standard Model, or how dark matter fits into the quantum structure of our universe.

Twenty years ago, physicists thought they had a pretty good idea how to detect the particles that make up the dark matter and to begin to measure their properties. The arguments based on the WIMP miracle seemed to make it clear that the discovery of these particles should be within the reach of the kinds of large underground detectors that we have since constructed— and possibly within the reach of the LHC as well. But as time passed, these signals have remained elusive. Perhaps one day soon, we will detect the particles that make up the dark matter and restore the WIMP paradigm to its preeminent status. But maybe not. Perhaps these experiments are telling us something else altogether. Perhaps the dark matter is not a WIMP at all. But if not a WIMP, then what?

When scientists make a prediction that does not pan out, they ultimately have to ask themselves what went wrong. Was there a flaw in the logic of their argument? Was something else discovered along the way that might have changed the expected outcome? Or might any of the assumptions that underpin the argument need to be revisited?

Back in chapter 7, I explained how the amount of dark matter that survives the Big Bang depends on how frequently these particles interact with ordinary forms of matter. From this calculation, we were led to think that the dark matter is likely to interact through the weak force or through some other unknown force that is roughly as powerful. Interactions of this kind were expected to be detectable. But as our experiments have continued to come up empty-handed year after year, we have been increasingly forced to consider the possibility that dark matter might not interact with ordinary forms of matter and energy nearly as much as we once imagined. A decade ago, we thought of dark matter as weakly interacting. Today, we are asking whether dark matter interacts with the visible world at all.

Behind the calculation that provides the foundation for the WIMP miracle, there are several important caveats, or assumptions. And although these assumptions all seem fairly likely to be true, they are receiving far more scrutiny these days than they ever used to.

Consider the transition that WIMPs are thought to have undergone during our universe's first moments. According to this standard picture of events, our universe originally contained a huge population of dark matter particles, which were constantly being created and destroyed in the ultra-hot bath of matter and energy. But only a fraction of a second later, there were far, far fewer of these particles left, and those that did remain had all but stopped interacting with one another. This description of events might turn out to be correct, but it relies on a key assumption about the nature of dark matter. Namely, it assumes that the dark matter interacts with other forms of matter often enough to be produced in abundance in the early universe. But what if that's not the case? It's possible that the dark matter is so feebly interacting that it was never produced in large quantities in the first place. If that were the case, then instead of the standard story—in which the bulk of the dark matter is destroyed, leaving behind a small remnant—the dark matter may have been populated only later, either gradually or suddenly, as cosmic history advanced.

There are good reasons why physicists didn't worry about this very much when they were originally considering the calculation behind the WIMP miracle. For one thing, if the dark matter interacts with ordinary matter through anything but the absolute feeblest of forces, then the very early universe would automatically be brimming with dark matter. Even a force a million times weaker than the weak force would do the job. But what if no such forces exist? Perhaps the dark matter simply does not interact through any of the known forces or with any of the known forms of matter or energy, other than through the effects of gravity. If this is the case, then the dark matter may have existed and evolved almost entirely independently of the rest of the matter and energy in our universe, opening up many possibilities that would have been unthinkable for an ordinary WIMP.

You could even imagine that the dark matter might not be alone, but might instead be only one of many forms of matter that almost never interact with any of the known particles. These elusive particles would constitute what physicists call a hidden sector, which could evolve and interact in any number of potentially complex ways. If hidden particles have no way to interact with ordinary matter, one might worry that they would survive the Big Bang in vast numbers, wildly exceeding the abundance of dark matter measured in our universe today. But through interactions between multiple kinds of hidden matter, it may have been possible for the abundance of dark matter to have been sufficiently depleted. In fact, it's quite easy for particle physicists to come up with viable hidden sector models that behave in this way.

Among the particles that make up a hidden sector, there could be forces and interactions that we have never observed, simply because they are not experienced by any of the forms of matter or energy that we currently know about. If we had never observed any quarks or gluons—or particles made up of quarks and gluons—then we would have no way of knowing about the strong force. If the dark matter is part of a large and complex hidden sector, it might experience forces that cause it to become bound to other forms of hidden matter and energy, forming what might be called dark nuclei or dark atoms. One day, we could even discover something like a periodic table of the hidden sector elements. The possibilities are almost endless.

When cosmologists calculate how much dark matter was produced in and survived the first moments after the Big Bang, they must assume something about the nature of this substance and its interactions. But they also have to make some assumptions about the universe itself. In particular, the results of this calculation depend critically on how fast our universe was expanding during its first moments.

In order to calculate the rate at which space expands, physicists use Einstein's general theory of relativity. More specifically, we use the equations that were derived from Einstein's theory by Alexander Friedmann in 1922. The inputs to these equations are pretty simple and come down to just two things. The first of these is the geometry of space, which in the case of our universe is flat, or at least very close to flat. The second is the quantity of energy present. Today, for example, the average energy density throughout our universe is just slightly higher than 10^{-23} grams— the equivalent of about 6.3 protons—per cubic meter. This number includes the contributions from atoms and other ordinary matter, as well as that from dark matter and dark energy. Using Friedmann's equations, this density translates to an expansion rate of about an inch per second for every light year that separates two points in space. Two galaxies separated by a billion light years, for example, will be carried away from each other at a rate of about a billion inches per second. At this rate, it will be 100 million years before the distance separating these two galaxies increases by even 1 percent. In human terms, our modern universe is evolving at an incredibly slow rate.

But this was not always the case. When our universe was young, it was very dense and very hot, and this means that it contained enormous quantities of energy. Take our universe as it was one second after the Big Bang, for example. The energy density of the sea of particles that filled all of space at that time was the equivalent of an incredible 2×10^{15} grams per cubic meter—or about a billion times the density of Earth. At this density, space expands wildly faster than it does today. Instead of a lethargic inch per second per light year, space was pulling apart at an instantly recognizable rate. If our universe were expanding right now as it was then, I would see my coffee cup— now just a few centimeters away—drifting away from me at a rate of about a meter per second. Objects across the table would be rushing away at a speed comparable to a major league fastball. And everything more distant than about 10,000 kilometers would be moving away faster than the speed of light—entirely lost from our view.

When we perform this kind of calculation, we take into account all of the known forms of matter and energy, including all of the kinds of particles that we have ever observed at the LHC and other accelerators. But as we have already discussed, the LHC has its blind spots. There may very well exist forms of matter that we don't know about, but that were abundant in the early universe. If this is the case, then the universe may have expanded quite differently from the way in which we currently calculate. And if the early universe was expanding either faster or slower than we now expect, this would change how the dark matter particles interacted, as well as how much of this substance survived.

The range of possibilities for how our universe may have expanded and evolved during its first second is enormous. Unknown forms of matter and energy may have increased the rate of expansion, but it's possible that far stranger things occurred during these first moments as well. In performing these kinds of calculations, cosmologists generally assume that space expanded and cooled at a gradually slowing rate during this period of time. But perhaps these earliest moments did not play out so simply or steadily. There are good reasons for us to speculate that dramatic events and transitions may have taken place during early cosmic history—the kinds of events and transitions that may have significantly impacted the origins of dark matter. For example, we know that in order for the particles that make up the atoms found in our universe to have survived the heat of the Big Bang, the total quantity of matter must have somehow come to exceed that of antimatter. According to Sakharov's third condition, this is only possible if our universe underwent a sudden and violent transformation during its early history.

Perhaps our universe experienced a brief and sudden burst of expansion, or underwent a dramatic phase transition at some point during its first second. Alternatively, there may have been a population of particles that decayed, heating the universe and altering its evolution. The possibilities abound. Such events could have dramatically impacted how the dark matter was formed and interacted during the first moments of our universe's history. If we were to learn one day that such an event really did take place, this would almost certainly change our expectations about the nature of dark matter and about the kinds of experiments that we would need to carry out in order to detect it. It might even explain why the dark matter has remained so elusive for so long.

Over the course of this chapter, I've described some very strange ideas. And with so many possibilities to consider, it may seem like I'm engaging in entirely unconstrained speculation, with little or no empirical evidence upon which to rely. To some extent this is true. There are many things about the early universe that we do not currently understand and have little ability to test, either through astrophysical observations or with experiments here on Earth. And this situation opens the door to speculation—some of which can be wild. Of all of the ideas I've put forth in this chapter, perhaps none seems as fantastical as the possibility that our universe once exploded in a sudden burst of expansion. But as it turns out, this particular possibility is also very likely to be true.

Over the past several decades, a considerable body of evidence has accumulated in support of the conclusion that our early universe did, in fact, experience a brief period of hyperfast expansion—an event known as cosmic inflation. Although inflation lasted only a little longer than a millionth of a billionth of a billionth of a billionth of a second, it left our universe radically transformed and utterly unrecognizable. To further thicken this plot, cosmologists have more recently learned that the expansion rate of our universe is not slowing down as had long been expected, but instead has begun to accelerate over the past several billion years. Although there is still much we don't know, it appears that our universe is in the process of entering a new era of rapid expansion.

The universe we live in and experience has little in common with either of these extreme eras. Our world, it seems, is in a moment of calm between two incredible storms. All indications are that our universe both emerged from, and will likely return to, a very different cosmic state.

10

A Flash in Time

Truth is the daughter of time.

—JOHANNES KEPLER

THE UNIVERSE WE SEE around us today is a universe filled with matter and radiation. These substances take on a multitude of different forms, from stars and planets to galaxies embedded in the halos of dark matter. Although these materials may be diverse, they have one thing in common: they were all forged in the heat of the Big Bang. And ever since, they have been slowly diluted by the gradual expansion of space.

But our universe was not always so tranquil. Although we find ourselves living in an era of steady and continuous change, the majority of modern cosmologists are convinced that our universe once, in its first trillionth of a trillionth of a second, underwent a violent and explosive burst of almost instantaneous expansion. During this era—known as cosmic inflation—space expanded so rapidly that essentially everything was pulled apart from everything else at speeds far greater than the speed of light. Under these conditions, any two particles would be ripped away from each other long before they could possibly interact, leaving each and every quantum particle in isolation. Inflation was a very lonely era. The very nature of space and time guaranteed it.

And then—as quickly as it began—cosmic inflation came to an end, leaving in its place a steadily expanding universe. Although this new state was still unimaginably hot and compact, it was not nearly as alien as the conditions found during the epoch of inflation. In a sense, one can think of the end of inflation as the true beginning of the universe that we live in.

Although most modern cosmologists are reasonably confident that inflation, or something like it, took place during our universe's first moments, this idea is a relative newcomer to scientific discourse. Nothing like inflation had ever been discussed among the first generations of scientists to study the Big Bang. Individuals including Georges Lemaître, George Gamow, and Ralph Alpher had each recognized that our universe was expanding and that it had emerged over billions of years from a hot and dense state. But nothing they knew about our universe led them to suspect that anything as strange as inflation had once taken place.

As recently as the early 1960s, the Big Bang was still widely perceived as a speculative and unlikely theory, residing somewhere on the outer fringes of respectable science. But within only a decade or so, the observational evidence in support of this cosmological paradigm had become overwhelming. The discovery and subsequent measurements of the cosmic microwave background placed the Big Bang Theory on solid empirical footing, only to be further strengthened by a long series of other measurements and observations. By the end of the 1970s, this body of observational evidence was sufficiently compelling to convince the bulk of the scientific community that our universe did, in fact, emerge from the hot and dense state that we call the Big Bang.

It was also around this time that cosmologists began to notice some unexplained problems with this theory. First of all, observations had revealed that the geometry of our universe is roughly flat, with no large degree of either positive or negative curvature. In other words, the rules of Euclidean geometry apply—parallel lines remain parallel as you follow them, and the angles of triangles add up to 180 degrees. This might not seem like a problem, but it was perplexing to many of the scientists studying the Big Bang. As a universe slowly expands, it should become more curved and less flat as time goes on. So in order for our universe to be approximately flat today, it would have had to have been almost *exactly* flat when it was young—to within 1 part in 10^{60}. Finding out that our universe is so flat is like stumbling upon a piece of stone in your backyard that is flat to within the width of an atom. Such an incredible degree of flatness simply couldn't have resulted by chance; it requires some sort of explanation. But at the time, cosmologists simply didn't have one.

Even more problematic was the incredible uniformity of the cosmic microwave background. In all directions of the sky, the temperature of this radiation is almost exactly 2.7255 degrees above absolute zero. The hottest and coldest points in the sky vary from this number only at a level of about 1 part in 100,000. This fact perplexed cosmologists, who began to ask how this could be possible. If you collected a set of 1,000 stopwatches from 1,000 different locations and found that they were all perfectly synchronized to within a fraction of a millisecond, you would quite reasonably assume that they must have been coordinated at some point in the past. But when it came to the cosmic microwave background, no one understood how this radiation—originating from very different parts of our universe—could have come to be so identical. Everything we knew about our universe at the time said that this should have been impossible.

To better understand this situation, imagine that I wanted to coordinate in some way with beings in a distant galaxy, such as one of those Edwin Hubble studied in the 1920s. Since nothing can move through space faster than the speed of light, the swiftest way to do this would be to send a beam of light in their direction with instructions. Let's say the galaxy in question is 3 million light years away from us—you might then think that this beam of light would reach its destination 3 million years from now. But we have to remember that the space between the galaxy and us is expanding, causing the galaxy to be moving away from us—and us away from it—at a speed of about 150,000 miles per hour. As a result, any light leaving today will take a bit longer than 3 million years—about 3,000,670 years—to reach its target galaxy.

Next imagine a more distant galaxy—one that is hundreds of billions of light years away from us instead of only a few million. Not surprisingly, the space between this galaxy and us is expanding much more quickly than in the last example. But there is more than a quantitative difference between this example and the last. The space that exists between two places separated by hundreds of billions of light years is not only growing at a speed that is very fast, but faster than the speed of light. Although nothing can move *through space* faster than the speed of light, space itself doesn't have to abide by this limitation. Distant points in space can expand apart from one another at any speed. As a consequence, no light—or any other kind of signal that we might choose to send—will *ever* reach this galaxy. And nothing they send in our direction will ever reach us. Because of the persistent expansion of space, we can never communicate with, see, or otherwise experience such a distant place. It is completely cut off from our observation, as well as from our influence.

In effect, the expansion of our universe causes all points in space to be surrounded by an impenetrable horizon, beyond which nothing can be observed and no communications can reach. In the current epoch, our cosmic horizon is the surface of a sphere with a radius of about 46.5 billion light years, with us at the center.[1] This sphere is, for all practical purposes, our universe. Its volume contains all of the space that we can see, study, witness, experience, communicate with, influence, or be influenced by. The boundary of this sphere is entirely impenetrable and forever will be—at least as long as the expansion rate of our universe doesn't begin to slow down. Things beyond this distance are not merely obscured from our vision, but are utterly disconnected from us. All of space beyond this distance is forever lost from our world.

Just as they are now, all points in space throughout cosmic history were surrounded by cosmic horizons. The size of these horizons, however, has been steadily changing over the course of time. Because our universe was expanding more rapidly when it was young, its horizons were much smaller then than

---

1. You might think that since only 13.8 billion years or so has passed since the Big Bang, nothing could be 46.5 billion light years away from us. Because space is expanding, however, an object that released its light 13 billion years ago and whose light is reaching us today would currently be much farther than 13 billion light years away from us. So, there is really no contradiction here to worry about.

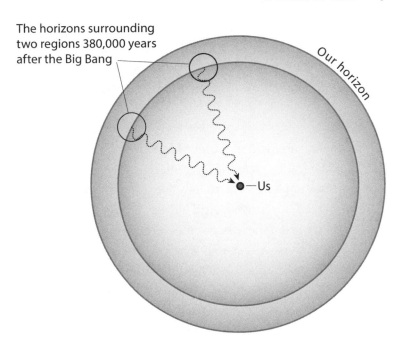

The horizons surrounding
two regions 380,000 years
after the Big Bang

Our horizon

Us

Because of the expansion of space, we are surrounded by a cosmic horizon. Beyond this horizon, we cannot observe or otherwise interact with anything. When the cosmic microwave background was emitted 380,000 years after the Big Bang, each point in space was surrounded by a much smaller horizon. This raises the question of how the temperatures of these regions came to be so nearly identical, even though they had no way of interacting with each other.

they are now. During the formation of the cosmic microwave background—about 380,000 years after the Big Bang—each point in space was surrounded by a horizon about a million light years in radius. From our vantage point today, a region of space this size appears to be about one degree in diameter. This means that the background of microwave radiation that we observe across our sky does not originate from one causally connected region, but from many thousands of independent regions that had absolutely no way of influencing each other. But without any way to interact with each other, how did these regions come to have temperatures that are so nearly identical? To return to my earlier analogy, it is as if we found a collection of thousands of nearly perfectly synchronized stopwatches, with no explanation for how they came to be that way in the first place.

By the 1970s, the Big Bang Theory had become clearly established as the leading scientific paradigm for the evolution and origin of our universe, and yet the flatness and horizon problems remained unresolved. And although most cosmologists may not have given these issues much thought during this period of time, they represented major blemishes on what was otherwise a very successful theory. In hindsight, we now recognize that these problems were telling us something about the earliest moments of our universe's history—an era that was not nearly as simple as most cosmologists at the time had imagined.

Somewhat surprisingly, the solution to these problems came not from a cosmologist, but from a young and relatively unknown particle physicist named Alan Guth. After hearing about the flatness and horizon problems faced by Big Bang cosmology, Guth began to think about scenarios in which our universe may have experienced a period of ultra-fast expansion during its first moments—the first theory of cosmic inflation. If our universe expanded quickly enough during this era, Guth reasoned, then this could explain why our modern universe is so flat. Whereas a universe that is filled with ordinary matter and radiation becomes only more curved as it expands, a universe undergoing inflation becomes steadily more flat—like blowing air into a balloon. On top of this, regions that were entirely disconnected when the cosmic microwave background was forming would have been very much connected to each other prior to inflation. By expanding so quickly, inflation took a relatively

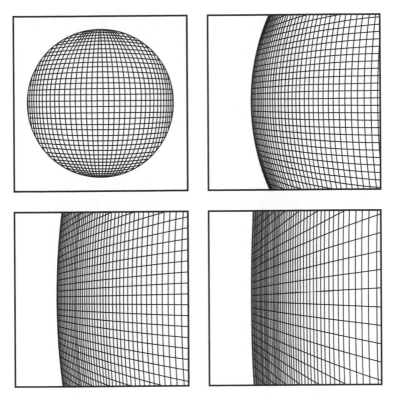

As the universe expands during inflation, the curvature of space is reduced, much as the surface of a balloon becomes increasingly flat as it is inflated.

small but nearly uniform region of space and stretched it into a volume that is larger than our entire observable universe. This means that every point in space that we can see today was once in contact with every other, providing us with a possible explanation for how the temperature of the cosmic microwave background could be so uniform.

But even if these problems could potentially be addressed by hypothesizing a period of ultra-fast expansion, this realization on its own was not yet a solution. The more important, and perhaps more interesting, question was how and why something like inflation could have taken place. What could have caused our universe to undergo such an incredible and explosive burst of growth?

The seeds of the idea of inflation can be traced back almost as far as the very origins of cosmology itself. Shortly after Einstein presented the first cosmological solution to the equations of general relativity in 1917, scientists began to discover other possibilities. The strangest of these was identified by the Dutch astronomer and mathematician Willem de Sitter.

In some ways, de Sitter's cosmological solution had a lot in common with Einstein's. Both solutions were derived assuming a universe that was homogeneous, and both included the presence of Einstein's famous cosmological term. But de Sitter's solution differed from Einstein's in a crucial way. It contained no matter. The universe that de Sitter had imagined was, in this sense, entirely empty.

Many at the time regarded de Sitter's solution as little more than a mathematical curiosity. After all, our universe isn't empty—matter plays a central role in our world. In the solution presented by Einstein, the tendency of a matter-filled universe to contract was pitted against the cosmological term, leading to a static balance. In contrast, the evolution of de Sitter's universe is driven entirely by the effects of the cosmological term. Without any matter to push against, the cosmological term in de Sitter's universe causes space to expand exponentially, ceaselessly increasing in volume. In fact, if the size of the cosmological term is large enough, de Sitter's universe looks a lot like one that is undergoing a period of cosmic inflation.

In the paragraphs above, as well as in chapter 3, I've talked about the cosmological term as if it were purely a mathematical entity. But when a physicist includes a mathematical construct such as this in their theory, it must also have a physical interpretation. If we want to take something like de Sitter's cosmological solution seriously within the context of inflation, it begs the question of what this cosmological term actually signifies. What does the cosmological term physically represent?

Put simply, the cosmological term—whether de Sitter's or Einstein's—describes the energy that is somehow stored in empty space. In a universe with a cosmological term, every cubic meter of space—at every place and at every time—contains a fixed quantity of energy, on top of any matter or anything else that it might contain. In other words, if you were to somehow remove every last atom, photon, and everything else from a volume of space, that space would still contain energy in the form of the cosmological term. This energy is a property of the vacuum of space itself.

In Guth's original proposal, our universe had a large cosmological term during a brief period of time shortly after the Big Bang. This cosmological term caused our universe to behave just as de Sitter had described, expanding exponentially. This part of the idea is relatively straightforward. The trick, it turns out, is in figuring out how this inflationary era came to an end. Guth imagined that the state of our universe during inflation was somehow unstable. After remaining in this state for a time, our universe underwent a spontaneous transition into a more stable configuration, without a large cosmological term, causing inflation to come to an end. But this proposal had a problem of its own: the end of inflation left the universe with almost nothing in it. The exponential expansion of space had diluted the densities of matter and radiation to virtually zero, with no way for them to be replenished. Whereas we know that our universe was once unimaginably hot and dense, Guth's was instead cold and empty.

Soon enough, however, other cosmologists proposed ways in which inflation could come to an end while leaving the universe full of matter and radiation. Scientists including Andrei Linde, Andy Albrecht, and Paul Steinhardt each came up with ways in which inflation could end gradually, allowing the energy that had once been stored in the vacuum of space to be transformed into a vast number of hot particles. In this way, the energy that powered inflation also heated the universe as this era came to a close.

Inflation offers a powerful explanation for how our universe came to be so flat and so uniform. But are we really sure that it is the right explanation? And even if it is, how would we know?

Since it was originally proposed, inflation has steadily increased in popularity among cosmologists. Although there are a few who say that they are absolutely certain that inflation really occurred, the vast majority agree that it is by far the best hypothesis we have devised to address the flatness and horizon problems. For one thing, a number of the predictions that follow from inflation have turned out to be true. In particular, most versions of inflation predict a fairly specific pattern for the temperature variations in the cosmic microwave background—a pattern that has been beautifully confirmed by measurements made over the past twenty-five years or so.

More specifically, inflation predicts that the distributions of ordinary matter, dark matter, and radiation in the early universe should all trace each other—that is, where there is a high density of one of these substances, there will also be a high density of the others. This is something that cosmologists have, in fact, observed to be the case. Furthermore, inflation predicts that the primordial variations in density should be approximately—but not perfectly—"scale invariant," meaning that the pattern of density variations across a given region will look roughly the same as the pattern you would find if you were to zoom in on the image. In this sense, inflation predicts that our universe should be similar to what mathematicians call a fractal. Cosmologists quantify this property using something known as the spectral index of primordial fluctuations, which most theories of inflation predict should be just a little less than 1 (where 1 is the value for a perfectly scale-invariant pattern). The most recent measurements of the cosmic microwave background place the actual value of this quantity at about 0.965—entirely consistent with what inflation had long predicted.

But despite these observational successes, some cosmologists remain uncertain that inflation actually took place. The evidence is compelling, but it is not entirely ironclad. And even if inflation did occur, the fact is that we still don't know much about the details of this era. Physicists have proposed many models, but it's simply very hard to tell which, if any, provides the correct description of the inflationary epoch. The kinds of direct and detailed evidence of this era that we would like to possess have long ago been buried beneath thick layers of energy and time. Much like other facets of our universe's earliest moments, inflation has been very difficult to definitively confirm or refute.

One of the most prominent and outspoken critics of inflation is also one of its creators—the theoretical cosmologist Paul Steinhardt. In the early 1980s, Steinhardt played an important role in developing the first workable theories of inflation, and if inflation were to be decisively confirmed one day, he would be a likely contender for the Nobel Prize. These days, however, you are unlikely to hear a good word about inflation from Steinhardt's lips. When he isn't focusing his attention on his other scientific interests—he tells a great story about hunting quasicrystals on the Kamchatka Peninsula—Steinhardt is often working to develop alternatives to inflation or trying to convince his colleagues that inflation is much less likely to have taken place than they might think. Among the true believers in inflation, Steinhardt is an apostate.

The objections raised by Steinhardt and his allies come in many forms. For one thing, in light of recent measurements of the cosmic microwave background, they argue that the only theories of inflation that can accommodate the data are those in which certain characteristics take on extremely specific and thus improbable values. In their view, inflation essentially solves the problem that our universe appears to be in a state whose flatness and uniformity would otherwise be extremely unlikely by proposing an era of inflation with characteristics that are themselves perhaps equally unlikely.

When it comes to inflation, however, what is and is not seen as likely can often depend on the eye of the beholder. The kinds of inflation that best agree with the current data are generally those models in which the rapid expansion of space was driven by a single quantum field that evolved very gradually—like a ball rolling across a very slightly tilted plateau, rather than falling down a steep hill. In order to produce a universe like the one we live in, the topology of this plateau has to have some rather specific features—in particular, its surface has to be broad and very flat. Is this unlikely? It's hard to say. Some prominent experts say yes, while others say no. It may be just an intrinsically subjective question. In any case, it is fair to say that the inflation's final judgment is yet to come.

If our universe never underwent an era of inflation, one is forced to ask how it came to be both so flat and so uniform. If we were to reject inflation one day, we would still need answers to these perplexing questions. If not inflation, then what left our universe in such a strange and specific state?

To date, no proposed alternatives to inflation have garnered much support among cosmologists. To the overwhelming majority of us, inflation remains our best hypothesis for how our universe came to be so flat and so uniform. But this doesn't mean that no other ideas are being considered. Steinhardt and some of his supporters, for example, have been exploring for years the possibility that instead of a burst of inflation, our universe may have begun with a bounce.

To understand what I mean by this, imagine that at some point in the distant future the expansion rate of our universe were to gradually slow down, come to a halt, and reverse. Although we have no indications of this, and nothing we know about our universe or its expansion history leads us to anticipate such a reversal, we can't entirely rule out the possibility that something like this might happen one day. Once in a contracting state, such a universe would become increasingly dense and hot, undergoing something like our universe's history in reverse. Ultimately, this kind of contraction would culminate in what is known as a Big Crunch.

What exactly would happen if a universe were to undergo a Big Crunch is something of an open question. If we take the equations of general relativity at face value, such a universe would collapse into a singular point of space and time—perhaps ending just as it began at the moment of the Big Bang itself. But the fact is, we don't have any good reason to believe that we can trust these equations under such extreme circumstances. Regardless of whether we are thinking about the Big Bang or the Big Crunch, we are dealing with temperatures and densities that are many trillions of times higher than anything we have ever studied at accelerators or in any other environment. It is entirely plausible that the equations of general relativity break down and fail to accurately describe nature at such incredibly high densities and temperatures. In the vicinity of a neutron star or black hole, for example, we can't rely on the equations of Newtonian gravity—under such extreme conditions, the effects of general relativity become too important to ignore. Perhaps in environments such as the Big Bang or the Big Crunch, there is yet another theory—one that we have not even begun to think about yet—that takes precedence over general relativity. Although it's hard to know what such a theory might look like, it is not hard to imagine that such a theory could predict a very different ending to a universe's contracting phase. Just maybe, instead of ending in a singular point of space and time, a Big Crunch might somehow reverse itself, not bringing the universe to an end, but instead starting everything all over again in what would look very much like a new Big Bang.

If a Big Bounce enables a universe to transition between contracting and expanding states, this opens the possibility that our universe may not have had a beginning and may never have an end. Instead, cosmic history could consist of an eternally repeating sequence of expansion, contraction, and bounce. Many variants of such cyclic cosmological models have been proposed over the years—even Einstein himself considered such possibilities in the 1920s and 1930s. More recent models sometimes involve extra dimensions of space, beyond the three that we know and experience. Within these theories, the forms of matter that we experience are confined to what is known as a brane—a three-dimensional space that is, in reality, a subset of the greater multidimensional space. Our world appears to us to be three-dimensional solely because electrons, photons, and the other known forms of matter can travel only within the brane's limits. The brane itself, however, can move through the full dimensionality of space. Within this kind of picture, something that looks like a Big Crunch followed by a Big Bang could occur whenever our brane collides with another. According to Steinhardt and his allies, these models are at least as likely to be true as those predicted within the paradigm of inflation.

It is fair to say that most modern cosmologists have not been impressed by the cyclic models proposed so far. Although such a theory might one day offer an explanation for how our universe came to be so uniform and so flat, the existing models are more like sketches than fully fleshed-out physical theories. They leave many questions unanswered and suffer from a wide range of theoretical problems. Furthermore, these models simply cannot explain many of the aspects of our universe that we see all around us—such as the detailed pattern of temperatures observed in the cosmic microwave background. Perhaps these problems will all be solved in time. But so far, cyclic cosmologies have simply not been as successful as inflation at explaining our universe as we observe it.

Like most healthy debates in science, this one will ultimately be answered not by theoretical considerations, but by observations and measurements. Over the course of the next decade, even higher-precision measurements of the cosmic microwave background will put many of the proposed inflationary scenarios to the test, allowing us to narrow down the range of possibilities and hopefully converge toward a true description of this strange and mysterious epoch.

Over the next five to ten years, scientists will scrutinize the cosmic microwave background with unprecedented precision. In addition to measuring things like the spectral index of primordial fluctuations in greater detail, these observations will offer much greater sensitivity to many of the subtle features that inflation may have imprinted into this radiation. Inflationary models involving multiple quantum fields, for example, often predict the presence of features known as non-Gaussianities. And even more exciting are the patterns known as B-mode polarization, which are predicted to have been generated by rippling waves of space and time as inflation was coming to an end.

In most models, the size of these so-called B-modes is directly related to the amount of energy that was present during inflation. Thus by measuring these B-modes, we could essentially measure the energy density of the vacuum during the inflationary epoch. Only a handful of years ago—in March of 2014—the cosmological community was thrilled when scientists operating the BICEP2 Telescope announced that they had for the first time detected the B-modes associated with inflation. If accurate, this measurement would have indicated that the energy density present during inflation was roughly the equivalent of what could be studied with an accelerator a trillion times more powerful than the LHC—a staggeringly high energy density indeed. Within only a few months, however, it had become clear that BICEP2's signal was not cosmological in origin—it was instead the result of emission from ordinary galactic dust. To this day, we have not yet observed the cosmological B-modes that would reveal to us so much about the nature of inflation.

As inflation tore our universe apart, ripples began to form in the very fabric of space and time. It is these moving ripples—known as gravitational waves—that created the B-modes that cosmologists are searching for in the detailed patterns of the cosmic microwave background. But in addition to their subtle impact on this background of radiation, these waves should themselves have survived the conditions of the early universe. Just as the heat of the Big Bang left our universe filled with a background of cosmic light, the epoch of inflation may have created a background of gravitational waves that still ripple throughout all of space and time today. But whereas the cosmic microwave background has enabled us to learn about our universe as it was 380,000 years after the Big Bang, this background of gravitational waves carries information about a far more primordial epoch. If we could detect and measure these ripples of space and time, they would provide us with a glimpse of our universe less than a trillionth of a trillionth of a second after the Big Bang.

We are all familiar with any number of kinds of waves, such as those associated with water or sound. In some ways, gravitational waves are quite similar to these more familiar types of waves, while in other ways they are very different. Whereas water and sound waves propagate by distorting the water, air, or other medium through which they travel, gravitational waves distort the very geometry of space itself. Gravitational waves are moving and periodic variations in the curvature of space.

Before Einstein, there was no reason to imagine that anything like a gravitational wave could exist. Within the Newtonian paradigm, it was thought that the effects of gravity moved instantaneously across space, leaving no room for any wave-like behavior. But gravity is very different according to Einstein. According to relativity, nothing can move faster than the speed of light, including gravity itself. There exist solutions to the equations of general relativity that describe how the effects of gravity move through space, including some that correspond to propagating waves of oscillating space and time. These gravitational waves travel at the speed of light and temporarily change the geometry of space as they pass through it, causing that space to expand and contract, stretching back and forth in time.

It's not hard to make a gravitational wave. They are created almost every time that anything is accelerated. The Earth itself releases a steady flux of rippling distortions into the surrounding space as it moves in its orbit around the Sun. But these gravitational waves are so incredibly weak that they are utterly imperceptible to any plausible experiment. To detect gravitational waves, physicists and astronomers have to rely on far more extreme events—the mergers of neutron stars and black holes.

Throughout most of history, the practice of astronomy was limited to what people could see with their own eyes. Since Galileo, astronomers have been using telescopes to see fainter and more distant objects, but even these observations were long limited to the kinds of light that happened to fall within the narrow range of wavelengths perceptible to human vision. But by the middle of the twentieth century, astronomers had begun to employ new technologies capable of detecting many other kinds of light. These new kinds of telescopes could perceive not only visible light, but also ultraviolet and infrared radiation, as well as X-rays, gamma rays, radio waves, and microwaves. Modern astronomy is sometimes not even limited to light. In 1987, for example, neutrinos were detected from a relatively nearby supernova explosion. There are hundreds of scientists today who call themselves neutrino astronomers.

Gravitational-wave astronomy began on September 14, 2015, when gravitational waves were detected for the first time by a large collaboration of scientists operating what is known as the LIGO observatory. These particular gravitational waves were created when two enormous black holes—about twenty-nine and thirty-six times as massive as the Sun—crashed into each other and merged about 1.3 billion years ago. Over billions of years, these black holes had been steadily releasing gravitational waves into space, causing them to lose energy and move closer and closer to one other. As their orbits tightened, they spun faster and released more energy in the form of gravitational waves. Finally, these two black holes spiraled together and collided at nearly half of the speed of light, ultimately forming a single and even larger black hole. In was in the last second leading up to their merger that this pair of black holes released the flash of gravitational waves that LIGO detected.

By using words like "flash," I might be giving you a false impression that these gravitational waves strongly distort the space that they pass through, or that they are otherwise easy to detect. Nothing could be further from the truth. As a gravitational wave passes through the Solar System, it stretches space only to the slightest degree. But the LIGO observatory is extremely sensitive to such changes. Over a length of 4 kilometers, its instruments are capable of detecting changes as small as 10^{-19} meters—about one ten-thousandth the size of a proton.

Over the coming years and decades, we expect LIGO to detect gravitational waves from dozens, if not hundreds, of black hole and neutron star mergers. As I'm writing this chapter, LIGO has reported the detection of eleven such events—ten mergers of black holes, and one of neutron stars. But by the time you're reading this book, there is a good chance that this list will have become substantially longer. This is a very rapidly moving field. I think it's fair to say that gravitational wave astronomy is the single hottest topic in all of science today.

LIGO has performed spectacularly and has thoroughly fascinated scientists and fans of science all over the world. But we will never learn much about inflation from this observatory. If we ever want to detect the background of gravitational waves that was formed in our universe's earliest moments, we will need to embark upon an even more ambitious experimental program—a program to study gravitational waves using an observatory deployed in space.

LIGO is designed to detect gravitational waves that ripple through space with a frequency of about 50 to 2,000 oscillations per second. But the gravitational waves from inflation are expected to be of a much lower frequency, making them impossible to detect on Earth. Fortunately, plans are underway to deploy a constellation of three spacecraft in orbit around the Sun, which would collectively function as a gravitational wave detector called LISA. Whereas LIGO is 4 kilometers in length, LISA's components will extend millions of kilometers across space. After its launch sometime in the 2030s, LISA will be sensitive to a wide range of phenomena that LIGO simply cannot detect. This includes the gravitational waves from supermassive black holes, with masses of millions or even billions of times that of the Sun. LISA is also expected to detect tens of thousands of mergers between black holes, neutron stars, and white dwarf stars. And last, but certainly not least, it is hoped that LISA will directly detect the background of gravitational waves that was left over from the early universe. This includes not only the gravitational waves created during inflation, but also those that may have been generated in phase transitions or other dramatic events that took place in the first fraction of a second after the Big Bang. With LISA, we will peer into our universe's early history in a way that we have never been able to before. I honestly don't know what we might learn. But with these new gravitational eyes, we will be able to penetrate through many of the layers of energy and time that have long obscured our universe's earliest moments. As the history of science has repeatedly shown, when you look at something with new eyes, you are very likely to find something that you have never seen before—and perhaps something that you had not expected or even imagined.

Despite the considerable successes of the Big Bang Theory as it was originally envisioned, it could not explain why our universe is so flat or how it came to be so uniform. To address these problems, modern cosmologists have reimagined de Sitter's solution to Einstein's equations, proposing that there was an exotic epoch within our universe's first moments during which space expanded in an explosive burst. But as strange as our universe's inflationary era may have been, it no longer seems to be entirely unique. Over the past two decades, a long series of detailed observations have revealed that our modern universe is once again expanding at a steadily accelerating rate. In other words, over the past several billion years, our universe has been transitioning into something that resembles a new—although much milder—era of cosmic inflation.

In order for our universe to be expanding faster today than it was yesterday, the vacuum of space throughout our universe must contain a small density of energy—the equivalent of only a few protons per cubic meter. The density of this so-called dark energy is so small that we can't detect its effects on Earth or even within our Solar System. But averaged across much larger volumes of space, dark energy dominates our universe— making up about 69 percent of the total energy, with the remaining portions consisting of dark matter (26 percent) and atoms (5 percent), along with trace amounts of neutrinos and light. The fact is, we don't really understand what this dark energy is. And that means that we don't understand why our universe's expansion rate is accelerating.

Our best observations to date suggest that the dark energy behaves just like Einstein's cosmological term—uniform in density throughout both space and time. In particular, as our universe expands, the density of dark energy remains the same—at least to within the limits of our measurements. In other words, the dark energy does not seem to be diluted by the expansion of space. Cosmologists sometimes quantify this behavior in terms of what is known as the dark energy equation of state. A perfectly uniform and nondiluting cosmological term corresponds to a value of −1 for this quantity. The most recent measurements tell us that the true value of the dark energy equation of state is within 5 percent or so of this value. And as I am writing this chapter, there are rumors that an even tighter measurement is about to be announced by a group of scientists known as the Dark Energy Survey, which has been conducting a systematic series of observations of distant supernovae, galaxies, and galaxy clusters. So far, these and other observations suggest that dark energy is uniform and nondiluting—just like Einstein's cosmological term.

In some ways, our modern era of accelerating expansion has some features in common with the epoch of cosmic inflation. But these two eras also differ in a number of critical ways. First of all, we estimate that the vacuum energy density during inflation was an incredible 100 orders of magnitude or so larger than that of dark energy today. This explains why our current era is so calm, while inflation was so exceptionally violent. After all, whereas inflation instantly tore every last particle apart from every other and utterly transformed our universe in only a minuscule fraction of a second, the accelerating expansion of space being driven by dark energy today was only noticed after careful observations across vast distances.

And second, whereas inflation came to an end, there are no signs of dark energy disappearing from our universe. If this trend persists indefinitely, dark energy will become only more prominent as the densities of matter and radiation are diluted by the expansion of space. From this perspective, the far future of our universe appears to be one in which matter and radiation are increasingly rare and only dark energy exists in large quantities. As our universe evolves, it will come to look more and more like the one described nearly a century ago by de Sitter—devoid of matter and entirely dominated by the cosmological term.

Few cosmologists doubt that dark energy exists—the evidence is overwhelming. This discovery, however, has been very challenging for scientists to explain or understand. Almost no one had expected our universe's vacuum to contain this amount of energy, and there is no consensus about what its presence tells us about our universe. One idea has been proposed, however, which could explain why dark energy exists in our universe in the quantity observed. This idea, which is very controversial among cosmologists, involves not only the twin mysteries of dark energy and inflation but also the existence of a multitude of universes. To the proponents of this idea, our universe may be the way it is because of life itself.

Endless Worlds Most Beautiful

I often think that we are like the carp swimming contentedly in that pond. We live out our lives in our own pond, confident that our universe consists of only the familiar and the visible. We smugly refuse to admit that parallel universes or dimensions can exist next to ours, just beyond our grasp.

—MICHIO KAKU

ON A SERIES of clear nights starting in 1609, an Italian mathematician at the University of Padua witnessed several things that no human being had ever seen before. Although he was not the inventor of the device that made this observation possible— the first telescopes had originated in the Netherlands about a year earlier—the mathematician had succeeded in building one of his own based solely on descriptions he had heard. Before that night, the telescope—or spyglass, as it was then more often known—was a tool reserved for terrestrial applications, such as observing approaching ships or distant armies. In the fall of 1609, however, Galileo Galilei decided that he would put this tool to another use, and for the first time try his hand at astronomy.

For his first astronomical target, Galileo selected the brightest object in the night sky—the Earth's Moon. Night after night, Galileo watched and studied the surface of this object. With the power of his telescope, he saw things that few at the time would have expected to see. The surface of the Moon, he discovered, was covered by deep valleys, high mountains, and flat plains. To his surprise, and in contradiction to all conventional wisdom, what Galileo witnessed was "rough and uneven, and just like the surface of the Earth itself."[1]

In addition to the Moon's terrain, Galileo's telescope enabled him to see many previously invisible stars and to discover four moons in orbit around Jupiter. This later observation was particularly important, as it clearly established for the first time that not all celestial bodies moved in orbits around the Earth. It was this, combined with his observations of the phases of Venus in late 1610, which convinced Galileo that the Earth-centric description of the universe could not possibly be correct. With his own telescope-aided eyes, he witnessed moons move around Jupiter, and Venus move around the Sun. Despite what had been believed for so long, the Earth was not the center of our universe.

Galileo's observations further revealed something both surprising and profound about the planets and moons themselves. He learned that these objects were—at least in some ways—similar to our own world. Just like the Earth, the Moon

---

1. Despite Galileo often being credited with the first telescopic observations of the Moon, similar observations were made months earlier by the English astronomer and mathematician Thomas Harriot. Harriot, however, never published his drawings of the Moon's surface and is far more well known for his other accomplishments, such as introducing the potato to Great Britain and Ireland.

had mountains and valleys, Jupiter had moons of its own, and Venus orbited the Sun—all of which flew in the face of the accepted teachings of the day. Through the strong influence of Christian theology and Aristotelian philosophy, it was widely taken as an unquestionable truth that the Moon, Sun, planets, and stars were fundamentally different from the world of our own experience. The greatest minds of medieval Europe had reached a consensus that the celestial bodies were perfect realms of pure and unchanging geometrical unity, quite unlike anything found on our chaotic world of disordered, elemental material. And just as a truly perfect sphere has no bumps or scratches, the surface of the Moon and planets—it was contended—most certainly contained no mountains or other such defects.

From a modern perspective, it can seem hard to believe that this opinion could have been ubiquitous throughout medieval Europe. After all, even with the naked eye, there are visibly darker and brighter portions of the Moon's surface—the "Man in the Moon" being the best-known example. But the philosophers and theologians of Galileo's day were thoroughly dedicated to Aristotle's cosmology and were willing to go to great lengths to explain away these facts. Among other things, they argued that different parts of the Moon's surface absorb and reflect sunlight differently, giving the false impression of varying terrain. The possibility that the Moon was anything but a perfect and unchanging sphere was inconceivable. But the light rushing toward Galileo from the Moon's surface had a very different story to tell. Aristotle's idealistic vision of the cosmos was not to survive the scrutiny of the telescope.

I've often wondered what went through Galileo's mind as he was making these incredible discoveries. Some individuals might have seen these newly revealed facts about the Moon's terrain as little more than novelties or trivia. I suspect, however, that Galileo recognized the much greater significance of what he had seen. He had, after all, just learned that the Earth is not unique. For the first time, we knew that other worlds exist.

Today, we take it for granted that the Earth is not the only planet in our universe. In addition to the eight in our own Solar System, astronomers over the past twenty years have discovered thousands of planets in orbits around other stars. The more we search for such planets, the more common they appear to be. It now seems that planets are the rule, rather than the exception—most stars do, in fact, have planets of their own. Around the approximately 100 billion stars within our galaxy, there are almost certainly hundreds of billions, or even trillions, of planets. It gives the imagination almost free rein to contemplate the range of possible realities that could exist across this collection of seemingly countless worlds.

This chapter is not, however, about worlds in the form of planets. Developments in cosmology have provided us with increasingly compelling reasons to think that other universes exist. Many of these universes are likely to be very different from our own, while others could be quite similar. In any case, I am talking about worlds that are entirely disconnected from our own realm of existence—worlds that fall beyond the reach of even the most powerful telescopes.

When I argue that it is likely that there are universes beyond the one that we know and experience, I often encounter considerable skepticism, even outright opposition. But in at least some respects, we know that other universes simply have to exist. Consider the well-established fact that our universe is expanding. Because of this expansion, every point in space is surrounded by a cosmic horizon, beyond which one could

never travel—even in principle. We will never be able to observe or interact with anything that lies beyond our horizon,
and someone beyond our horizon could never observe or interact with us. As far as I'm concerned, a place that one could
absolutely never observe, interact with, or travel to is not in the
same universe. Although there are many different definitions of
this word, I think of a universe as a region of space in which
things could in principle interact. In this sense, our universe is
only a small piece of a greater landscape of expanding space,
consisting of a multitude of causally disconnected worlds.

Cosmologists have long wondered how far space extends
beyond our horizon, and how many different disconnected regions might be present within that larger space. The fact is, we
really have no idea. All indications are that space is truly expansive and contains a vast number of such regions—the overwhelming majority of space appears to reside well beyond our
cosmic horizon. But that is about all that we know. As far as
Einstein's equations are concerned, it is possible that space simply extends outward forever in all directions. It is also possible,
however, that space eventually wraps around on itself. In a
space with this type of geometry, traveling far enough in one
direction will eventually lead you back to where you started—
like walking around the surface of the Earth, or flying off of the
edge of the screen in the classic video game "Asteroids." No one
can say for certain whether the space beyond our horizon extends forever, or whether instead it ultimately wraps around on
itself or in some other way comes to an end.

If space is truly infinite, the implications are staggering.
Within an infinite expanse of space, it would be hard to see any
reason why there would not be an infinite number of galaxies,
stars, and planets, and even an infinite number of intelligent or
conscious beings, scattered throughout this limitless volume.

That is the thing about infinity: it takes things that are otherwise very unlikely and makes them all inevitable. If you roll a million dice, the odds of them all coming up sixes is tiny— about 1 chance in 10 to the power of 788,151. You could roll dice all day long for many lifetimes without ever rolling anything that unlikely. But you could not live an infinite number of lives without seeing this happen an infinite number of times—any finite number, no matter how small, becomes infinite when multiplied by infinity. Similarly, within an infinite amount of space, there is an infinite number of universes, and within those universes, all things possible are realized—no matter how unlikely they seem to be.

The universe as we know it can be thought of as a collection of atoms and other particles located at specific places at specific times. For any given distant universe, there is some very, very small—but not exactly zero—probability that all of its atoms and other particles are oriented in almost exactly the same way that they are in our world. In other words, within an infinite space, there are inevitably an infinite number of universes that are indistinguishable from our own. These worlds contain a star that is nearly identical to the Sun, which is orbited by a planet that is nearly identical to the Earth, which contains upon it people who are nearly identical to you and me. If space as we know it extends forever, this conclusion is inevitable. All things and all events that are possible, no matter how unlikely, will exist and will occur within this greater collection of space. Such is the nature of infinity.

The expansion of space divides it into a number of causally disconnected regions. For all intents and purposes, these regions are not part of a single universe; rather, each is a universe of its own. Each point in space is surrounded by an impenetrable cosmic horizon, the size of which is determined by how fast space is expanding. The faster that space is expanding, the closer this cosmic horizon will be to the point that it surrounds. During eras of accelerating expansion—such as our current era—space is continuously being divided into a larger number of disconnected universes. And while this is taking place to some degree in our universe today, there was a point in cosmic history during which the accelerating expansion of space was far more dramatic. Nothing created a multitude of universes faster than cosmic inflation.

During the epoch of inflation, space expanded at an absolutely staggering rate, tearing space and everything in it apart. No two objects—even elementary particles—remained close enough to one another for long enough to interact. Two objects separated by the width of an atom at the beginning of inflation were trillions of miles apart from one another by the time it was over—only a minuscule fraction of a second later. Inflation took regions of space that had once been neighbors and forever disconnected them from each other. So utterly complete was this act of sequestration, that these regions became more than merely distant. Inflation left them in entirely different universes.

A small piece of the space that emerged from inflation went on to form our universe. But there is good reason to think that everything we can see in our sky represents only the smallest tip of the cosmic iceberg. During inflation, countless pieces of space were stretched into newly formed universes, populating a greater multiverse of disconnected worlds. And despite the fact that we have no way to observe this panoply of universes, there is every reason to suspect that it does, in fact, exist.

In some fraction of these universes, matter and energy could take on forms that are the same or at least similar to those we find in our world—such as atoms and light. A universe that contains the same basic building blocks of matter as ours and has the same underlying laws of physics is likely to look a lot like our world. This doesn't mean that there was a Battle of Hastings in 1066 in these universes, or that Richard Nixon was elected president of the United States in them; the specific circumstances of history would play out differently in each one. But it does mean that stars, planets, and galaxies would form in such universes and that nuclear fusion in those stars would lead to the formation of many of the chemical species that we find in our Solar System. And with the right chemistry, present under the right circumstances, it becomes possible for complex things—like battles and presidents—to emerge.

Not all universes within the greater multiverse need be so similar to our own, however. In some, the laws of nature could be subtly different—or very different—from those we observe. Patches of space separated by inflation could evolve in such a way that they come to support the existence of different kinds of matter and forces. Much as the Earth's flora and fauna varies from place to place, it is possible that different regions of the multiverse could be dictated by a diversity of physical laws.

If the laws of physics in our universe were even slightly different from what they are, the characteristics of our world could be dramatically altered. For example, if the strength of the electromagnetic force were a mere 4 percent weaker than it is in our world—or if the amount of electric charge carried by protons and electrons were only slightly smaller—protons in stars would be able to bind together, releasing huge amounts of energy and causing the Sun to immediately explode. On the other hand, if the electromagnetic force were much stronger than it is, then carbon atoms would be unstable—making life as we know it impossible. If protons were a mere 0.2 percent heavier than they are, or if neutrons were 0.2 percent lighter, free protons would decay into neutrons, instead of the other way around, leaving the world devoid of stable atoms. In these and many other ways, the characteristics of our universe seem to be remarkably well suited for the emergence of life.

Some regions within the multiverse are likely to be very different from the world that we know and experience. New forces and new forms of matter may populate many of these realms of existence. In some, there might be more—or fewer—than three dimensions of space. Many worlds may be utterly unlike anything we can imagine. At present, we are very far from anything like a complete understanding of this subject. Physicists have only recently begun to contemplate seriously just how the laws of physics that rule our universe came to be the way they are and how they might be different in others.

Throughout the history of science, physicists have been working to develop and improve our understanding of our world and how it functions. Little by little, however, we are beginning to think within a greater context. Instead of merely studying and describing the phenomena we observe, we are asking ourselves questions of *why*. Whereas physicists have traditionally asked questions such as "How do electrons interact with one another?" or "What are protons made of?," they are increasingly thinking about *why* these things are the way they are. Not merely how nature is, but why it is the way that it is. And could it be different elsewhere?

I ended the previous chapter by suggesting that it might be possible to explain the mystery of dark energy by invoking a very large number of universes, and that this might also somehow be connected to the event of cosmic inflation. But what do these seemingly very different ideas and events have to do with each other? How could the number of universes that exist beyond our cosmic horizon possibly have any impact on the quantity of dark energy that we find to be present in our world?

Dark energy is the energy present in empty space—the energy of the vacuum. In principle, physicists think they know how one would go about calculating the density of this energy. Because of the quantum nature of our world, it is possible for particles to be spontaneously created, only to disappear a moment later. This process is taking place constantly, throughout all of space. At this very moment, particles—pairs of electrons and positrons, for example—are coming into and out of existence all around you, being created and destroyed by nothing more than the potentiality of the vacuum itself. These particles may be short lived, but they collectively contribute to the energy density of so-called empty space. The problem is, when we calculate the expected energy density of the vacuum, we get a number that is roughly 10^{120} times larger than the measured abundance of dark energy. It seems that these kinds of processes do not generate nearly as much dark energy as the math says we should expect.

Let's pause for a minute to consider just how wrong this calculation appears to be. I've done a lot of physics problems in my life, and I doubt that I've ever gotten an answer that was this different from the right one. I'm not trying to brag here—I've gotten plenty of wrong answers throughout my education and career. But it is actually very hard to get an answer *this wrong*. For example, if one somehow managed to miscalculate the mass of a proton by the same factor that we've misestimated the density of dark energy, they would get a value of about 10^{93} kilograms—or about 10^{63} times the mass of the Sun. Even a very poor student rarely gets an answer that is that stupendously incorrect.

To be clear, the problem with this calculation isn't that someone made a simple mistake—no one forgot to carry the 1. But perhaps there might be some kind of conceptual problem with the way that contemporary physicists are thinking about this calculation. What this problem might be, however, is anyone's guess. Decades of intense scrutiny have not led to much progress or clarity. If there really is a problem with the way that we are performing this calculation, we haven't a clue what that problem is.

In 1987, years before the discovery of dark energy, the physicist Steven Weinberg proposed a very different approach and solution to this problem. Weinberg is one of the most respected physicists alive today and is one of the original architects of what particle physicists call the Standard Model. He is not known for being reckless or imprudent with his ideas, and he is no provocateur. He is considered by his peers to be a very serious thinker and a first-class scientist. What Weinberg proposed in 1987, however, made many of his colleagues uncomfortable, and some downright angry. In a paper, he argued that our universe's vacuum energy density could be explained if we take into account the fact that we are living in it. All one needs for this to be the case is for there to exist an incredibly large number of universes. The idea of the modern multiverse had been born.

Let's walk through the argument behind Weinberg's proposal. The amount of dark energy in a given universe determines how that universe expands and evolves, at least in the long run. Whereas matter and radiation are diluted by the expansion of space, the density of dark energy remains constant. In our universe, for example, dark energy began to dominate the overall energy density about 9.8 billion years after the Big Bang, at which point the expansion rate began to accelerate. But the time at which this accelerated expansion kicks in depends on how much dark energy is present. In a universe with much more dark energy than ours, this acceleration will begin much earlier and will have a much more dramatic effect than it does in our universe.

Imagine, for example, a universe that starts out just like ours, except for the fact that it contains a million times more dark energy. In this universe, the expansion rate begins to accelerate only 20 million years or so after the Big Bang. Up until that point in time, this universe looks a lot like ours—its light nuclear elements have formed, and a similar mixture of dark matter, hydrogen, and helium fills all of space. But once the expansion of space begins to accelerate, the history of this universe becomes very different from that of our own. In our universe, gravity caused dark matter to clump together to form halos, which went on to form galaxies and led ultimately to the formation of stars and planets. But in a universe with a million times more dark energy, space expands too quickly, and these things never get a chance to take place. And without stars, none of the heavier elements would ever have been created. In a universe without stars or planets—or even carbon, oxygen, silicon, or iron—it's hard to imagine how anything as complex as life would ever emerge. A universe with so much dark energy is a universe that could never contain life.

If we assume that the way in which modern physicists have calculated the energy density of the vacuum is essentially correct, then the overwhelming majority of the universes in the greater multiverse will contain vast quantities of dark energy—approximately 10^{120} times more than the amount found in ours. But the result of this calculation depends on the details of the inputs—things like the kinds of particles that exist and the strengths of their interactions. If any of these quantities vary from universe to universe, then we should expect some of the universes to contain more dark energy while others contain less. And although they would be exceedingly rare, some universes should even have as little dark energy as we find in our world.

It might seem strange at first glance that we would happen to find ourselves living in such a highly unlikely universe. But as Weinberg pointed out, we have to be careful to ask the right questions. Instead of asking what fraction of universes within the greater multiverse is like ours, we need to ask ourselves where a typical living being within the multiverse is likely to find him- or herself.

Throughout the overwhelming majority of the multiverse, dark energy is so abundant as to make life impossible—no living observers will ever find themselves in such a place. In fact, any universe with more than about ten times as much dark energy as there is in ours will almost certainly be devoid of life. Across the multiverse, life is very rare indeed—but it is not impossible. Taking into account both the distribution of universes and the likelihood that things like planets and stars will form within a given universe, Weinberg and others have calculated how much dark energy typical observers should find to be present in their universe. Intriguingly, they find that the most common places for life to emerge are those corners of the multiverse in which there is roughly as much dark energy as there actually is in our world. In this sense, the existence of the multiverse resolves the long-standing mystery of dark energy.

Arguments such as Weinberg's are based on what is known as the anthropic principle, and it's fair to say that this is a controversial idea among cosmologists. At its core, this principle says only that any observer must exist somewhere that it is possible for an observer to exist. That hardly sounds objectionable. But in practice, one often runs into a variety of technical issues when trying to apply anthropic reasoning.

In Weinberg's calculation, for example, he has to assume that the density of dark energy varies from universe to universe and with a distribution that peaks strongly at very large values, with only a small tail that extends to the value that we observe in our world. Based on what we know about quantum field theory, these assumptions are entirely plausible, and perhaps even likely. But we don't know this for sure, and it's not clear how we ever would. And this brings us to the other kind of objection that is often raised to applications of the anthropic principle. Some argue that it's not even science.

The twentieth-century philosopher Karl Popper is most famous for his work on what is known as the demarcation problem, or the problem of how to distinguish truly scientific endeavors from those of pseudoscience. Most people think they can recognize science when they see it, but unless you've thought a lot about this issue, it can be pretty difficult to come up with a set of criteria that is satisfied by everything that we think of as science, while excluding things like astrology and moon landing conspiracy theories. Popper's solution was to insist that in order for a theory to be considered scientific, it had to offer predictions that could—at least in principle—be falsified. A theory that could never be disproven is not a scientific theory.

Critics of the multiverse and the anthropic principle often appeal to Popper's criteria of falsifiability. After all, there is no observation we could ever make that would definitively show that no other universes exist. It does not seem to be a falsifiable hypothesis. I'm of the opinion, however, that this is an overly simplistic view of the problem. After all, many of the philosophers who followed Popper rejected his criteria of falsifiability, and some even proposed other ways to distinguish science from pseudoscience. For example, it's been argued—and I agree— that what is really essential in order for a theory to be scientific is that some future information, such as observations or measurements, could plausibly cause a reasonable person to become either more or less confident of its validity. This is similar to Popper's criteria of falsifiability, while being less restrictive and more flexible.

Does the anthropic explanation for the observed density of dark energy count as being scientific according to this criteria? As I see it, it clearly does. After all, Weinberg wrote his paper on this subject in 1987—years before the presence of dark energy was observationally discovered. In his paper, he predicted that if his plausible-seeming assumptions were correct, then we should expect the energy density of our universe's vacuum to be about the value that it was later measured to be. And while those later measurements did not prove that the multiverse exists, it certainly made it seem more likely. That, in my opinion, is why the anthropic explanation for dark energy is a scientific one—even if we're not yet certain whether or not it is actually true.

If our universe does not exist in isolation, but is instead part of a greater multiverse, this would raise a series of new and interesting questions. In particular, this shift in perspective might change how we think about the origin of our world. Throughout this book, I've used the phrase "the Big Bang" to refer to the exceedingly hot and dense state that our universe emerged from some 13.8 billion years ago. Many people, including many scientists, think of the Big Bang as the beginning of our universe—or at least as the beginning of our universe as we know it. But if one takes a more global view within the context of the cosmic inflation, the Big Bang starts to look less like an event or a beginning, and more like an ongoing process. From this perspective, the origin of what we call our universe may have been the result of a perpetual mechanism rather than an isolated occurrence.

Before cosmologists proposed inflation, there were good reasons to think of the Big Bang as a singular event, in which space and time themselves came into existence. In the late 1960s, Stephen Hawking and Roger Penrose provided considerable support for this view by deriving mathematical proofs for what are known as the singularity theorems. Among other things, these proofs used the properties of general relativity to demonstrate that any universe filled with things like matter and radiation will be bounded in both space and time. In other words, the nature of gravity itself seems to guarantee that time must have had a beginning. According to these arguments, our universe began as a singularity, before which there was not only no matter, energy, or space, but also no such thing as time. From this perspective, there is simply no such thing as *before* the Big Bang.

The occurrence of inflation, however, radically changes how we think about the origin of our world. The singularity theorems derived by Hawking and Penrose rely on certain assumptions, known as energy conditions, which are clearly broken by inflation. If the expansion and evolution of our universe had been driven only by matter and radiation, the conclusions of these theorems would have been applicable. But in a universe that was dominated by the energy of the vacuum during inflation, there is nothing that guarantees that space and time began as a singularity—or even had a beginning at all.

From an inflationary perspective, the origin of our world is less like the spontaneous creation of space and time out of nothing and more like the birth of a new creature from a long line of ancestors. Imagine a region of inflating space—continuously expanding at an incredible and exponential rate. As the field that drives inflation evolves throughout this space, its characteristics change, and ultimately inflation comes to an end. The quantum nature of space and time, however, guarantees that this transition will not occur at all places at exactly the same time. Instead, one patch of space will stop inflating first, while others continue to rapidly expand. When inflation ends in a given region, that space becomes populated by a hot and dense plasma of energetic particles, which gradually cool as that space expands. In other words, each such patch of space becomes its own universe, potentially not so different from our own.

But what happens in those regions of space that did not stop inflating? Although one might guess that inflation will ultimately end everywhere, yielding a large but finite number of disconnected universes, even this understates what seems to be the most likely outcome. In most viable theories of inflation, the total amount of space that is actively inflating only *increases* as time moves forward. Although some regions cease to inflate, others grow more than fast enough to compensate. This means that as inflation goes on, new universes are born without limit, and inflation never ends. As space expands exponentially, inflation spawns new regions of the multiverse without ever stopping and without any end. In this sense, inflation appears likely to be eternal.

If inflation never ends, it is also possible that it never had a beginning. It may just go on forever, in both directions of time. But time itself within this context is a slippery and counterintuitive concept. According to Einstein's theory of special relativity, the order in which two different events take place can depend on the observer's frame of reference. While one person might witness one event taking place before another, a different person moving through space with a velocity near the speed of light could see the same two events occur in the opposite sequence. In the case of inflation, we are dealing with a large—perhaps infinite—number of causally disconnected spaces. No two observers in any two of these disconnected regions could ever check to find out which region existed first. In this sense, the order in which events occur in the multiverse is not always well defined.

Furthermore, if inflation really does continue forever, it could potentially go on to create every possible region of space. This means that if there were a region of space in which inflation originated—the first region of space—inflation might eventually create another one just like it, or at least arbitrarily similar to it. From this patch, inflation will seem to start all over again. Maybe that is how the first region got there in the first place? It would be as if one's great-great-great-great-great-great-grandchild were also their parent. When you have no way of assessing the order in which events take place, the sequence in which time unfolds does not necessarily need to be linear—and it certainly does not need to be intuitive to us.

In this chapter, we've waded deeper into speculation and scientifically controversial ideas than anywhere else in this book. As the evidence for inflation has become steadily more compelling, however, an increasing number of scientists have been willing to entertain and explore the possibility that our world is part of a multiverse, spawned by eternal inflation. But even this is not the limit of what is being imagined by cosmologists today. Well-motivated and promising theoretical frameworks such as string theory suggest that there may be an unimaginably large number of possible configurations in which the vacuum of space and time could exist. Each of these vacuum states corresponds not only to a different density of dark energy, but to different physical laws and to different forms for the matter and energy that occupy space.

And once we've gone this far, why stop here? Inflation may provide us with a mechanism through which a multiverse is likely to be generated, but it is also possible that universes exist without any spatial or temporal connection to our world at all. We can go on to ponder why one universe might exist, while another does not. Perhaps the logical possibility of a world guarantees its own existence? Or maybe there is some other cosmic principle that dictates which kinds of worlds do and do not exist within the greater multiverse. Today, we have no credible answers to questions such as these. Tomorrow, who knows?

Touching the Edge of Time

There should be no boundary to human endeavor.

—STEPHEN HAWKING

IT WAS ONLY a little over a century ago that the science of cosmology was born. Out of the radical ideas of Albert Einstein and the observational evidence that space is expanding, the modern cosmological paradigm emerged. And although the Big Bang Theory started out as a scientific pariah, the discovery of the cosmic microwave background in 1964 left little room for doubt that our universe did, in fact, evolve over billions of years from a hot and dense state. For the first time in history, human beings had begun to understand the origin of their universe.

Over the past fifty years, an army of astronomers and physicists have been actively scrutinizing our universe, learning ever more about its distant past. Today, we understand in some detail how our universe has expanded and evolved over most of its history. Any number of observations have confirmed the predictions of the Big Bang Theory to an incredible, and frankly unexpected, extent. The rate at which our universe has expanded over the past 13.8 billion years agrees with the equations written down by Alexander Friedmann almost 100 years ago, and measurements of the large-scale distribution of galaxies and galaxy clusters are indistinguishable from those predicted by the theory. And most impressive of all, the detailed pattern of temperature variations observed across the cosmic microwave background has been a treasure trove for cosmologists, revealing to us everything from the amount of atoms, dark matter, and neutrinos that are present in our universe, to the large-scale geometry of space itself.

Yet despite all of these successes of modern cosmology, there is no denying that many key mysteries remain unsolved, and we still know relatively little about our universe's earliest moments. In fact, several of our most recent discoveries have raised more questions than they have answered. Despite decades of effort, the nature of dark matter remains unknown, and the problem of dark energy seems nearly intractable—unless one is willing to indulge in the anthropic arguments that come with the existence of a multiverse. Although we have some reasonable guesses, we do not know how the particles that make up the atoms in our universe managed to survive the first moments of the Big Bang. And perhaps most elusive of all, we still know very little about cosmic inflation, how it played out, or how it came to an end—assuming that something like inflation in fact happened at all.

The history of science contains many examples of mysteries that were once as perplexing as these, but were eventually solved—in fact, this is the rule and not the exception. From this perspective, we may be tempted to feel secure that time and persistence will ultimately produce answers to all of cosmology's open questions. Betting against the progress of science—at least in the long run—is almost always a losing proposition.

But in the case of our universe and its origin, there are reasons to wonder how much progress we can make, and how much we will—or even can—one day learn. The horizon of our universe limits what we can observe and study, and thus limits the amount of information that we can access, even in principle. From this perspective, it is conceivable that there could exist questions about the Big Bang and the moments that followed that are simply impossible for us to empirically address. In this sense, there may one day be an end to the great line of human inquiry that we call the science of cosmology.

In the years ahead, there are many ways in which cosmology will continue to advance. It was only recently that astronomers observed the first gravitational waves, and this new way of viewing our universe is almost certain to lead to important developments in this field. More detailed measurements of our universe's large-scale structure and the evolution of its expansion rate will also be carried out in the years to come, revealing even more about the nature of dark energy. Particle accelerators such as the LHC and its possible successors will collide larger numbers of particles and at higher energies, enabling us to study the laws of physics as they functioned less than a trillionth of a second after the Big Bang. And experiments searching for the particles that make up the dark matter of our universe will continue their exponential growth in sensitivity. Although there is no way to be certain, I would be surprised if we don't conclusively discover the nature of dark matter at some point in the next decade or two.

In the history of cosmology, no observations or measurements have borne as much fruit as those of the cosmic microwave background. Providing us with a detailed snapshot of our universe 380,000 years after the Big Bang, this collection of photons represents by far the greatest body of information we have about our universe's youth. Among other things, if we had never observed the cosmic microwave background, we would know virtually nothing about inflation—in fact we would have had little reason to think that something like inflation took place at all. It was by scrutinizing the cosmic microwave background that we learned that our universe is so geometrically flat and uniform—precisely the features that inflation was originally proposed to explain. Furthermore, modern observations of this radiation have shown that the patterns of its temperature variations closely match what is predicted by most theories of inflation. These observations have convinced most cosmologists that something like inflation likely took place. That said, we still know very little about how this occurred, or how it came to an end.

Over the next decade or so, cosmologists will certainly continue to learn more about inflation by further examining the cosmic microwave background. New telescopes will be sensitive to the features of this radiation that we expect to be most informative about inflation and its aftermath. It is possible, if not likely, that these measurements will provide us with new and powerful insights into this earliest of epochs. But such endeavors also have a clearly defined limit. Quite soon, we will reach a point at which we will no longer be able to learn any more than we already have from the cosmic microwave background.

The problem with cosmology is that we have only one universe to study. If I want to learn about how atoms behave, I can perform experiments on any number of atoms and repeat them over and over again, collecting data and information without limit. But if I want to learn about how universes evolve, there is only one that I can observe, and I can't repeat the experiment to see if a universe might behave differently under different conditions. More specifically, our universe has only one cosmic microwave background, and it contains a finite amount of information. Once we have measured it to a sufficiently high degree of precision, there will be little to be gained by measuring it again. We are rapidly approaching the point at which we will have learned essentially everything that the cosmic microwave background has to tell us about our universe and its early history.

Fortunately, our universe contains a great deal of information in other forms that we have not yet been able to access. The cosmic microwave background was formed at a specific time in cosmic history—about 380,000 years after the Big Bang—when the contents of our universe transitioned from electrically charged nuclei and electrons to mostly neutral atoms. Ever since this transition, these neutral atoms have been emitting another kind of light. It is the glow of this cosmic atomic gas that will provide the next generation of cosmologists with their most powerful way of studying our universe and its first moments.

Hydrogen atoms radiate many different kinds of light, but cosmologists are most interested in the photons emitted as a result of what is known as the hyperfine transition. When an electron that is bound to a hydrogen atom spontaneously flips the direction of its spin, it emits a photon with a very specific frequency—1.42 gigahertz, corresponding to a wavelength of just over 21 centimeters. By studying the light that reaches Earth with this specific wavelength, astronomers can determine the distribution and temperature of the hydrogen gas that is present throughout the Milky Way and other parts of the nearby universe. But as these photons travel across even greater distances, they are stretched, or redshifted, by the expansion of space, causing their wavelengths to grow beyond the 21 centimeters that they were created with.

By studying this radiation at each given wavelength, cosmologists can study a different period of cosmic history. Consider, for example, the photons that were emitted by hydrogen gas 1 billion years ago. They reach us today with a wavelength of about 22.7 centimeters. In contrast, the photons emitted 8 billion years ago now take the form of 43-centimeter waves of light. The oldest of these photons—those released 13.8 billion years ago, at the time of the formation of the cosmic microwave background—have been redshifted to a wavelength of about 230 meters. So whereas the cosmic microwave background provides us with a snapshot of our universe at one particular moment in time, this background of redshifted 21-centimeter light provides us with something like a film of our universe's 13.8-billion-year history. By studying the cosmic microwave background, we learned that inflation likely took place. By measuring and scrutinizing the 21-centimeter light that has been emitted throughout cosmic history, we will strive to determine how this extraordinary epoch played out, came to an end, and led to the formation of our universe as we know it.

The redshifted 21-centimeter radiation that reaches Earth contains an incredible amount of information about our universe—far more than even the cosmic microwave background. This information is deeply hidden, however, and not at all easy to observe or extract. Although astronomers have been detecting 21-centimeter photons from the Milky Way since the 1950s, the more distant and interesting cosmological signal is much fainter and far more difficult to measure and study.

As I am writing this chapter, astronomers have begun to report their first observations of the redshifted 21-centimeter emission originating from the era known as the cosmic dawn. It was during this period that the first stars in our universe formed, a couple of hundred million years after the Big Bang. Within a decade, we hope to have maps of this emission across a wide range of wavelengths, spanning from the time of the cosmic dawn up to much more recent periods of our universe's history. To carry out this ambitious program, astronomers will not be able to rely on conventional telescopes, which are blind to light across this range of wavelengths. They will instead build and operate enormous arrays of carefully designed radio antennas, spanning areas as large as a kilometer or more in extent.

Although certainly promising, such endeavors face considerable challenges and limitations. For one thing, human beings produce an enormous amount of noise across this range of frequencies. Everything from cell phones to short-wave broadcasting makes it difficult for cosmologists to detect and study this signal—much as city lights tend to conceal the full beauty of the night sky. For this reason, astronomers are constructing their antennas in remote parts of South Africa, Canada, and the Australian Outback. Adding to these challenges, the Earth's ionosphere absorbs a large fraction of this radiation across much of the frequency range that is of most interest to cosmologists. Earth, it seems, is not a great place to study our universe in this way.

But if not from Earth, then from where? Human-made radio waves radiate in all directions from Earth, polluting the Solar System with a perpetual background of static noise. Any radio antennas that we might deploy on satellites or on the International Space Station would suffer from many of the same problems that we experience here on Earth. There is, however, one place that our cell phones and other radio signals cannot reach. There is no place in the inner Solar System that is more radio quiet than the far side of the Moon.

It seems likely—and even imperative—that scientists will ultimately turn to the Moon as the site for their most ambitious cosmic investigations. By deploying a few million simple radio antennas across a 100-kilometer region of the lunar surface, astronomers could go a long way toward extracting essentially everything that our universe has to tell us about its history, evolution, and origin. The secrets of cosmic inflation may indeed be waiting for us on the far side of Earth's only moon.

Even being optimistic, however, it will be decades before astronomers are able to deploy an array of radio antennas on the far side of the Moon. Although both NASA and the European Space Agency have started to make plans to build large scientific facilities on the lunar surface, they are at this stage more of a rough sketch than anything written in stone. It's hard to predict how things will pan out, but I wouldn't be surprised if we are doing cosmology on the Moon by sometime around the middle of the twenty-first century.

Combining measurements of the redshifted 21-centimeter emission with those of the cosmic microwave background, cosmologists will learn far more than they currently know about our universe's early history, and in particular the era of cosmic inflation. These endeavors will not only test inflation and put it on much stronger footing than it is today, but also fill in many of the gaps that currently exist in our understanding of this formative epoch. By revealing to us things like how much energy was contained in space during inflation and how this period of rapid expansion came to its close, these measurements will provide us with a new and powerful window into this essential and formative era of cosmic history.

So far in this book, I've said next to nothing about the period of time that preceded inflation. The reason for this is that we have essentially no observations or other data to tell us what our universe may have been like during its pre-inflationary era—the first 10^{-32} seconds or so after the Big Bang. But even without observations, it is not entirely fair to say that we know nothing about this period of time. By extrapolating toward higher temperatures and backward in time, we can make reasonably informed speculations about what our universe may have been like, even during these earliest of moments.

While inflation was taking place, our universe contained an enormous density of energy—it was this vacuum energy, after all, that drove space to expand at such a spectacular rate. This energy must have come from somewhere, and this means that before inflation began, space was filled with a dense sea of particles carrying enormous quantities of energy—likely corresponding to a temperature somewhere around 10^{29} degrees. Unfortunately, we have no way to observe or measure how matter and energy behave at such high temperatures; even the LHC cannot produce or collide particles with anywhere near this much energy. In fact, if we wanted to build a machine capable of accelerating particles to such speeds, it would have to be roughly the size of our Solar System. Needless to say, construction on that project will not be starting anytime soon.

All is not lost, however. By studying the laws of physics as they function at lower energies, we can extrapolate in an effort to infer aspects of how this early era may have played out. For example, physicists have long known that the strengths of the strong, weak, and electromagnetic forces each evolve with temperature. The strong force, for example, diminishes as temperatures rise, and thus it was considerably less powerful in the early universe than it is today. And although these three forces act with very different strengths in the present universe, our calculations indicate that they were all approximately equally powerful when the temperature of our universe was somewhere around 10^{28} degrees. Many physicists see this as an indication that these three forces are not as independent as they might seem, but are instead each a manifestation of a single, unified force. In fact, there are many features of the Standard Model that suggest that these three forces and the particles they act upon are all parts of a larger and more complete theory—what we call a Grand Unified Theory.

In some ways, Grand Unified Theories have a common ancestor in how physicists have come to understand electricity and magnetism. Prior to the nineteenth century, phenomena such as electric current and lightning were thought to have little or nothing to do with why compass needles point north. As far as the natural philosophers of the day were concerned, electricity and magnetism were separate facets of nature. But eventually it was noticed that whenever an electric field changed, a magnetic field was generated. In fact, we now understand that a magnetic field is nothing more than an electric field in motion. Electricity and magnetism cannot be separated or exist independently of one another, but are instead different aspects of the same electromagnetic force. If our universe is described by a Grand Unified Theory, as many physicists think is likely, then the strong, weak, and electromagnetic forces are deeply interconnected in a similar way.

According to most Grand Unified Theories, our universe experienced a major transition in its very early history as it expanded and cooled. In the very earliest of times—corresponding to temperatures well above 10^{28} degrees—the strong, weak, and electromagnetic forces functioned as a single unified force, acting upon an array of particle species that are far too massive for us to create in any plausible particle accelerator. During this time, the laws of physics that ruled over our universe were nothing like those we know or are able to study at colliders. But as space expanded and cooled, our universe underwent a phase transition that broke the unified force into three seemingly distinct sets of phenomena—the strong, weak, and electromagnetic forces. The farther we look back in time, the less recognizable we find our universe to be. But as time moved forward, space and the forms of energy that it contained rapidly transformed into something that increasingly resembled our world.

During the epoch of Grand Unification, our universe was a very alien place. But as we look back even farther and closer in time to the Big Bang itself, we find an era in which our universe was even more unrecognizable. During these first instants—the first 10^{-43} seconds—even the very nature of space and time was unlike anything that exists in our universe today. This was our universe at its most strange and mysterious. This was the era of quantum gravity.

Essentially all of modern physics is built upon two extraordinarily powerful theories: Einstein's general theory of relativity and the quantum theory of particles and fields. By any reasonable measure, each of these theories has been spectacularly successful. For a century physicists have been testing the limits of these theories, probing them with increasing precision in an effort to identify some situation or circumstance under which they might fail. But despite these considerable efforts, no such limit has ever been found. Under every condition that we have been able to observe and study, both general relativity and quantum theory have passed every test. Simply put, these theories provide us with a remarkably accurate and detailed description of our universe.

Yet we know that this cannot be the entire story. Physicists have long realized that one or both of these theories must somehow break down and fail to accurately describe nature under extreme circumstances. Despite the incredible successes of general relativity, it is not a quantum theory, and it is not compatible with other quantum theories.

Under ordinary circumstances, Einstein's theory predicts with incredible accuracy how objects will move through space, based on the geometry or curvature of that space. For any particle that we have ever observed in our universe, this procedure is straightforward and works spectacularly well. But we can at least imagine a particle that contains so much energy that its presence alone strongly distorts the space surrounding it. And if two or more such particles were in close proximity, they might even warp space to such a degree as to form a tiny black hole or otherwise alter the geometry of space beyond recognition. Making things even stranger, quantum particles are not generally located at only one place at one time. So under extreme temperatures and densities, quantum particles and the space surrounding them can be described only as a simultaneous collection of different geometrical configurations—what is sometimes called spacetime foam.

Just like any other quantum field, physicists imagine that this spacetime foam is somehow made up of individual pieces—or quanta. For a point of comparison, consider the electromagnetic force. Before the discovery of quantum physics, it was thought that this force could be described in terms of a continuous field extending across space. But at the quantum level, we now understand that the electromagnetic field is not a continuous thing, but is instead made up of a vast number of individual particles—photons. In a similar way, in order for the phenomenon of gravity to be compatible with the quantum nature of our universe, it must somehow be built out of discrete quanta—hypothetical particles known as gravitons. Some physicists even think that space and time themselves are probably made up of individual quanta as well.

These quantum aspects of space and time are imperceptible to our experiments and observations. Only at extremely high energies—10^{15} times greater than those studied at the LHC—does the geometry of space and time begin to exhibit its true quantum nature, and no particle with so much energy exists in our universe today. But energies such as these were once realized in nature, during the first 10^{-43} seconds after the Big Bang. During this era of quantum gravity, nothing—including space and time itself—was familiar or recognizable. Everything was different from anything you have ever imagined.

I do not want to leave you with the impression that quantum gravity is a well-understood problem. It certainly is not. And for this reason, most of what I am saying here about our universe's quantum gravity era could be partially, or even mostly, wrong. When one talks about the first 10^{-43} seconds after the Big Bang, one has little choice but to rely on a healthy dose of extrapolation and well-informed speculation.

When physicists first tried to extend the phenomenon of gravity as described by general relativity into a quantum theory—as we have successfully done with the other three forces—they quickly ran into some serious problems. In theories that describe the curvature of space in terms of a collection of individual gravitons—much as the electromagnetic field is a collection of photons—many of the calculations simply explode to infinity. In technical terms, these theories of quantum gravity are non-renormalizable—a clear indication that they are broken and cannot provide a true or logically self-consistent description of our universe.

In an attempt to overcome these problems, a huge amount of effort has gone into ideas such as string theory and loop quantum gravity. In string theory, fields are not made up of point-like quantum particles, but instead consist of extended quantum objects, such as strings, and sheets called membranes. Many physicists think that such a theory could one day provide a fully renormalizable and self-consistent description of quantum gravity. But as we understand string theory today, this is possible only if space and time are far more expansive than they appear to be. Strangely enough, string theory seems to require that our universe consists of either ten, eleven, or twenty-six dimensions of space and time.

At first glance, this might seem like a fatal flaw for the theory. The world of our experience is clearly four-dimensional—events occur at one point in time and at one location in three-dimensional space. But physicists have come up with a number of ways that other dimensions of space might exist, hidden from our direct experience. For example, it is possible that the other dimensions are simply too small for us to notice. Picture for a moment a direction in space that is perpendicular to each of the three standard spatial dimensions. And now imagine that this fourth dimension of space loops around on itself after only a very short distance—after moving only 10^{-35} meters in this direction, you return you to where you started, having circumnavigated the entire universe in this direction. Most quantum particles are simply too large to ever move through or otherwise experience this extra dimension. For them, it is as if it does not exist. From this perspective, it is possible that we really do live in a universe that is 10-, 11-, or 26-dimensional, but with most of these dimensions curled up into loops that are far too small for us to have noticed. The size of a quantum particle, however, depends on its energy—the more energy a particle has, the shorter its wavelength. For this reason, extremely energetic particles are also extremely small—perhaps small enough to travel through the full dimensionality of space.

Farther back in time, however, the particles that filled space possessed much greater quantities of energy and thus may have once been capable of traveling through dimensions of space that we do not experience today. In this sense, the dimensionality of our universe may have transitioned steadily downward as it expanded and cooled. If string theory turns out to be the right path to understanding quantum gravity, then it's probably the case that a series of such transitions took place throughout the first 10^{-43} seconds or so after the Big Bang. Although our universe became effectively four-dimensional (including time) rather early in its history, the true dimensionality of space and time was on full display during this brief moment that we call the era of quantum gravity.

To make things even stranger, string theorists have found that certain theories of gravity can be directly related to—or mapped onto—other theories, including those that describe the behavior of quantum particles and their interactions. In the 1980s, string theorists discovered five different versions of string theory: one 26-dimensional theory and four other theories requiring ten dimensions of space and time. They thought—or at least hoped—that one of these five theories would eventually be found to describe our universe, but were not sure which one. What they discovered in the early 1990s surprised everyone involved. Instead of being distinct and fully independent, it turns out that each of these five string theories provides a partial description of the same overarching

11-dimensional theory. Through certain mathematical operations, each of these theories can be transformed into any of the others. In other words, although each of these theories seemed very different on the surface, they all describe the same underlying phenomena.

This realization carried with it enormous implications for the nature of space and time. After all, if a 10-dimensional theory really describes the same underlying phenomena as a 26-dimensional theory, then should we think about spacetime as being 10-dimensional or 26-dimensional? In the quantum gravity era, there may not be a simple answer to this question. In this strangest of epochs, even what it means for something to be a dimension of space or time can be ambiguous. When we finally understand quantum gravity one day, it may turn out to be unclear whether we should think of our universe as having been 10- or 26-dimensional during its earliest instant. In a sense, we may even learn that it was both.

Over the past century, scientists have spectacularly resolved many of their most perplexing puzzles and questions. From the inner workings of the atom and the identity of DNA to the evolution of stars and the nature of light, the progress of science has exceeded all expectations. From this perspective, it is reasonable to imagine that many or most of the questions explored in this book will be answered in the years and decades ahead. Using history as a guide, we find that even the most difficult of scientific challenges have a way of being overcome.

But where does this all end? As we discover answers to the questions asked by science, new questions often arise. Throughout our history, we have discovered many new and powerful ways of conceptualizing our world, and this progress will almost certainly continue in the future. But centuries, millennia, or even millions of years from now, will there still be unanswered questions? One day, will we reach the end of science?

Personally, I'm skeptical that anyone—or any civilization, for that matter—will ever reach a true end of this grand project. For one thing, it is possible that the layers of understanding that are built into the structure of our universe simply continue on forever. A century ago, we replaced our view of gravity as a Newtonian force with Einstein's idea that gravity is the phenomenon associated with the geometry of space and time. And someday, I hope we will discover an even more powerful way to think about space and time—a theory of quantum gravity—which will supplant general relativity. Perhaps this kind of progress will simply continue on forever—theory superseding theory, without limit, each providing a deeper and more complete picture of our world, without our ever reaching any ultimate or final theory.

On the other hand, imagine that one day we do discover a theory that successfully predicts all known phenomena and answers all of the questions that we are capable of asking about our universe. Although we might be tempted to call this the final theory, there would be no way of knowing for certain. Even with a theory that successfully describes all observations to date, you could always ask yourself whether the first observation to disagree with the theory might be made tomorrow. From this perspective, the task of science can never really end. There will always be new and unanswered questions to ask.

But there seem to be some questions that science simply cannot answer—now or ever. Some of these are big questions—the kind that human beings of all times and cultures have asked themselves in some shape or form. Of all such questions, perhaps none has been pondered as many times by as many people as "Why is there something, instead of nothing?"

At first glance, this might seem like the kind of question that the science of cosmology would be well suited to answer. After all, among other things, cosmologists study how the universe began. But questions of *how* are fundamentally different from questions of *why*. We have all encountered a child who responds to every answer with that question, "Why?" If you're knowledgeable and a little lucky, you might be able to successfully answer the first few of these ripostes. But before long, you reach a point where you have no choice but to reply, "Because that's the way it is."

There may very well be facts about our world that are true just because they are—what philosophers call brute facts. Such statements would not require, and may not even have, any explanation. They are just true.

When it comes to why there is something rather than nothing, some physicists have argued that the quantum nature of reality sheds some light on this question. After all, discoveries in quantum physics have shown that objects can spontaneously come into existence—although generally for only a short time. And through inflation, even the tiniest piece of space can expand to become a full-fledged universe. Perhaps our universe itself came into being in this way.

But in my opinion, these kinds of explanations don't really address the question at hand. Just like a curious child, I can respond to such explanations with the question, Why? The quantum nature of reality might be able to explain how space and the energy and matter it contains came into existence—but where did the quantum nature of reality come from? And why does reality function in this way? Although it seems to be true that inflation can make a universe from *almost* nothing, the bridge from true nothingness remains beyond our powers of understanding. To me, these kinds of explanations leave us no closer to an answer than where we were when we started.

On the other hand, these answers are certainly no worse than those put forth by any number of philosophers and theologians throughout history. Aristotle concluded that a "prime mover" was required to put our universe into motion, and theologians have long argued that the existence of our universe can be explained only as an act of God. But these explanations are ultimately just as empty as appeals to quantum processes, and they get us no closer to answering the question at hand. Regardless of whether we rely on God or on the laws of quantum mechanics to address this question, there is no escape from further questions of why. You can push the brute fact as far back as you like, but every sequence of questions of the form "Why" ultimately ends with an answer of "Because."

When I ponder the state of cosmology today, I find myself alternating between two very different points of view. On the one hand, I see an incredibly successful scientific endeavor. Observations of our universe have made it clear that space has been expanding over billions of years, originating in the hot and dense state that we call the Big Bang. Over the past decades, this theory has been scrutinized and refined as new kinds of precision measurements enabled us to reconstruct the history of our universe in ever greater detail. And when we compare the results of different kinds of measurements—the expansion rate of our universe, the temperature patterns in the cosmic microwave background, the abundances of the various chemical elements, and the distribution of galaxies and other large-scale structures—we find stunning agreement. Each of these lines of evidence supports the conclusion that our universe expanded and evolved in just the way that the Big Bang Theory predicted. From this perspective, our universe appears to be remarkably comprehensible.

On the other hand, one doesn't have to look very far to find ways in which cosmologists have struggled—if not outright failed—to understand important facets of our universe. We know almost nothing about the dark matter and dark energy that together make up more than 95 percent of the total energy in existence today. We can only speculate about how the atoms that inhabit our world somehow managed to survive the heat of the early universe or how the era of cosmic inflation may have played out. And although each of these mysteries is fascinating in its own right, what really intrigues me is that each of them seems to point us toward the first moments that followed the Big Bang. In this sense, our inability to detect particles of dark matter, as well as the simple fact that atoms exist in our world today, suggests that the earliest moments of our universe's history may have included events, interactions, or forms of matter and energy that we still know nothing about. Cosmic inflation also took place during these earliest of times, raising far more questions than we currently have answers. These are our universe's greatest mysteries, and all indications are that their solutions lie in the first fraction of a second that followed the Big Bang.

Some scientists continue to view these puzzles as little more than loose ends, which they expect to be resolved through further investigation and scrutiny. But so far, these problems have proven to be remarkably stubborn and persistent. We designed and built the experiments that we thought would be required to detect the particles that make up the dark matter, and yet we found nothing. Nor has the LHC revealed anything that moves us closer to resolving any of these cosmic mysteries. Despite having measured the expansion history and large-scale structure of our universe in ever-increasing detail, we have not gained any substantively greater understanding of the nature of dark energy.

It is from this perspective that I find myself asking whether these cosmic mysteries might be the symptoms of something more important or expansive than a few loose ends. Perhaps these puzzles are not as unrelated as they might seem, but are instead each pointing us to a very different picture of our universe's earliest moments. When it comes to our understanding of our universe's origin, I sometimes find myself wondering: Is a revolution coming?

Scientific revolutions are often unexpected and go initially unnoticed by all but a few. Consider, for example, what it might have felt like in 1904. Never before had physics seemed to be on such solid footing. For over two centuries, physicists had been successfully applying Isaac Newton's laws of motion and gravity to problem after problem. And although progress made in the nineteenth century expanded our knowledge into areas such as electricity, magnetism, and heat, these newly understood aspects of our world were, in more ways than not, not so different from those Newton had described hundreds of years earlier. To the physicists of 1904, the world seemed very well understood. There was little reason to expect that a revolution was coming.

Similar to the situation faced by cosmologists today, however, there were a few problems in 1904 that physicists hadn't yet been able to address. The speed of light had been measured and found to be unexpectedly uniform—light always moves through space at the same speed, regardless of direction or time. Physicists had long thought that light traveled through space by distorting a medium, which they called the luminiferous aether.

But the uniformity of light's speed and the lack of any direct evidence for this substance had begun to present challenges to this notion. It was hard for physicists to understand why they had failed to detect the effects of the aether, but without it they had no cogent way of understanding the nature of light.

Physicists around this time were also confounded by the observed motion of the planet Mercury. For decades, astronomers had been aware that Mercury's orbit was slightly different from what was predicted by Newton's equations. Some even went as far as to suggest that an unknown planet, called Vulcan, might be perturbing Mercury's trajectory. But despite considerable effort, this so-called Vulcan had never been seen.

Perhaps the most conspicuous problem faced by physicists in 1904 was that they had no idea what powered the Sun. No known chemical or mechanical process could possibly generate enough energy to create so much sunlight for so long. As far as they were concerned, the Sun should have stopped shining a long time ago.

And lastly, to the physicists of 1904, the inner workings of the atom remained a total and utter mystery. Various chemical elements were known to emit and absorb light with specific spectral features, but physicists had no idea why. Even the very stability of atoms was difficult to understand from the perspective of Newtonian physics.

Few saw it coming, but in hindsight it's easy to recognize that these problems were heralds of a coming revolution in physics. And in 1905 a revolution did indeed come, ushered in by a young Albert Einstein and his new theory of relativity. We now know that the luminiferous aether does not exist and that there is no planet Vulcan. Instead, these fictions were symptoms of the underlying failure of Newtonian physics. Relativity beautifully solved and explained each of these mysteries, without any need for new planets or substances. Furthermore, when relativity was combined with the new theory of quantum physics, it became possible to explain the Sun's longevity, as well as the inner workings of atoms. These new theories even opened doors to new and previously unimagined lines of inquiry, including that of cosmology itself.

Scientific revolutions can profoundly transform how we see and understand our world. But radical change is never easy to see coming. There is probably no way to tell whether the mysteries faced by cosmologists today are the signs of a coming scientific revolution or merely the last few loose ends of an incredibly successful scientific endeavor. There is no question that we have made incredible progress in understanding our universe, its history, and its origin. But it is also undeniable that we are profoundly puzzled, especially when it comes to the first fraction of a second that followed the Big Bang. I have no doubt that these earliest moments hold incredible secrets, but our universe holds its secrets closely. It is up to us to coax those secrets from its grip, transforming them from mystery into discovery.

CREDITS

INDEX

Note: Page numbers in *italics* indicate figures.

act of God, 219
Albrecht, Andy, 168
"alpha-beta-gamma" paper, 53n1
Alpher, Ralph, 53, 58, 159
Andromeda Galaxy, 38, 91
Andromeda Nebula, 38
anthropic principle, 194
antimatter, matter and, 86, 92
antiquarks, 97, 98
Aristotle, 16, 17, 184, 219
Asimov, Isaac, 130
ATLAS detector, 66

baryogenesis, 96
Bekenstein, Jacob, 113
Bethe, Hans, 53n1
BICEP2 Telescope, 174
Big Bang, 5, 11, 13, 46–62, 160, 219; "alpha-beta-gamma" paper, 53n1; cosmic microwave background, 54, 60; cosmic radiation, temperature of, 54; cosmological redshift, 50; dark energy, 61; dark matter, 61;

expansion of space, *51*; galaxies, Hubble's discovery about, 47; heat of, creation of matter and antimatter in, 7; helium, quantities of in universe, 52; Holmdel Telescope, 57; key event in cosmic history, 56; nucleosynthesis, predictions of, 115; primordial state of, 45, 49; radiation left over from, 107; scientific community, key feature of, 47; stars, energy released by, 52; stellar nucleosynthesis, 52; supernovae, 52; Theory, 46, 50, 94, 164
Big Crunch, 171, 173
big questions, 217
bimetric theories, 112
black holes: crashing and merging of, 176; gravity of, 97, 136
B-mode polarization, 174
Bolyai, Janos, 18
bosons, 74. *See also specific types*
brane, 172
brute facts, 218
"Bullet Cluster," 113–114